Edward Micklethwaite Curr

Pure Saddle-Horses and How to Breed them in Australia

Vol. 1

Edward Micklethwaite Curr

Pure Saddle-Horses and How to Breed them in Australia
Vol. 1

ISBN/EAN: 9783744760904

Printed in Europe, USA, Canada, Australia, Japan

Cover: Foto ©berggeist007 / pixelio.de

More available books at **www.hansebooks.com**

PURE SADDLE-HORSES,

AND

HOW TO BREED THEM IN AUSTRALIA.

TOGETHER WITH A CONSIDERATION OF THE HISTORY
AND MERITS OF THE

ENGLISH, ARAB, ANDALUSIAN, & AUSTRALIAN
BREEDS OF HORSES.

BY

EDWARD M. CURR.

MELBOURNE:
WILSON & MACKINNON, PRINTERS AND PUBLISHERS,
COLLINS STREET EAST.
1863.

TO MY OLD FRIEND,

MYLES GERALD KEON,

FORMERLY OF THE FRENCH FOREIGN LEGION IN ALGERIA,

NOW CHIEF SECRETARY OF BERMUDAS,

THESE PAGES ARE INSCRIBED

WITH MANY FEELINGS OF AFFECTION, BY

THE AUTHOR.

PREFACE.

In offering this work to the public, none can feel better than the Author the necessity for asking some indulgence for entering on a theme which has already been treated of by so many. This becomes, apparently at least, the more needed by one to whom the subject is not professional, who has never backed, entered, or ridden a racer on the turf, and who yet ventures to differ in opinion with many received authorities on the subject of the horse in general, and to call in question doings and customs which have been considered beyond the reach of contradiction. But whilst differing radically from several able authors who have, with what has been considered excellent opportunities, written on this subject, I should be sorry to be thought to have undervalued their efforts or capabilities. So far, indeed, from doubting these, I cannot but wonder that they have done so much, hedged in by local prejudices and with an experience drawn from so very limited a field for observation as Great Britain; whose great error appears to have been their idea that whilst treating of the artificial and arbitrary specialties of Leicestershire and Epsom, they were retailing a cosmopolitan wisdom and dealing with the necessities of the Horse and the horseman of the world at large.

In considering this point, it has struck me as not unlikely that the undeniable superiority of the English horse on the turf, and the fame of the English hunter, placed side by side with the very degraded saddle-horse of the rest of Europe, has led to the dictatorial tone of Englishmen and English writers on this subject, and has encouraged them to claim a superiority for their saddle-horses over many others, which I have no hesitation in saying exists only in their imaginations, and would never stand the test of a trial.

But it is not merely the English estimate of the English horse that I complain of in connection with their treatises on the subject of the horse. Hitherto it appears there has been a great lack of knowledge shown by our writers of the horses of other countries. None of our authorities, indeed, appear to have troubled themselves with seeking for much information on a subject on which, notwithstanding, they write dogmatically enough. A few anecdotes or descriptions, drawn from books of travel, followed by some haphazard desultory remarks, seems fully to have exhausted in their minds all that is either useful or expected from them on the subject. Besides this, the testimony of travellers so quoted is never accepted farther than where it is found conveniently to dovetail into the preconceived ideas or prejudices of the writer. Any accounts of vigor, endurance, stoutness, or abstemiousness which the English author knows is not to be found in the English horse, he, unexamined, quietly puts aside as unworthy of belief, facetiously remarking on the well-known accuracy of travellers' tales, or the want of mile-stones in the desert. Nor is this little charlatanry very imprudent with readers but little disposed to set up rivals for their equine favorites. A little cold water on the subject of foreign horses has indeed gone a long way, and if to the Arab horse some "faint praise" has been meted out, it has only been out of consideration for the family honor as the progenitor of the famous thoroughbred.

But if the summing up of the untravelled English judge and the verdict of the unenquiring jury have always been in favor of the native and adverse to the foreigner, the case as yet may almost be said to have gone by default, for as yet the cause of the Eastern horse, which, as far as is consistent with truth, I mean to plead, has been but very feebly urged by such of our authors at least as treat professedly of the subject.

It is not, however, these apparently trifling

misrepresentations, nor the false deductions which are based upon them, but the ruinous prejudices which we have gradually accepted, that has induced me to take up my pen with the hope of being useful to my fellow-colonists.

It cannot but be allowed by the most enthusiastic believer in the virtues of the English horse, that however well qualified such writers as Youatt and Stonehenge may have been to treat of that animal, that they have advanced but very slender claims to entitle them to be held as authorities on any foreign horses, excepting as regards the latter author those of North America. A Frenchman, however good a judge he might be of horseflesh, who had seen a few packs of hounds throw off, or witnessed the Derby run, would not probably be deemed necessarily a very high authority by gentlemen of the turf; few of them, probably, would be much inclined to back his favorite for the Derby, though he might be considered quite *un jock*, and shake his head very knowingly, and write a sporting article or two on his return to his "native Tours." If the Frenchman's fitness for the post of a sporting oracle would be more than doubtful, is it not, may I ask, on the very same principle and with just that sort of information at best, that Youatt and Stonehenge have undertaken to treat of, compare, and judge the horses of the world. Many years of practical experience, joined to a natural aptness for the subject, alone fit perhaps one person out of every thousand who attempts the subject, to be held as a judge of these matters in his own country; but the shallowest smattering of anecdotical reading concerning foreign horses, quite unsupported by personal experience, seems to be all that is requisite to mature a knowledge and perfect appreciation of the despised "outsider." A traveller, who would be looked on as a fool at Ascot if he should back his own judgment in the betting ring, is referred to as a great authority on the Turkish horse, because he has past a fortnight at Constantinople or Bayakdere, and hazarded some remarks on an animal, which possibly no previous training had fitted him to estimate.

Of saddle horses in general, their merits, and the limits of their powers, I feel no hesitation in saying, that the English horseman, of only English experience, knows but little, and is a much less competent judge than the Australian bushman Nor is it singular that it should be so : for he wants opportunity, he is never in a position to arrive at a full knowledge of the animal. The English gentleman, if ever he requires a roadster, buys one under the advice of a *vet* or experienced groom, or perhaps commissions some one, in whose judgment he believes, to buy for him. If a colt, the breaker breaks him and trains him ; the groom feeds and looks after him ; the farrier shoes him ; and the veterinary surgeon prescribes for him when he ails. In the bush, on the contrary, we breed, and occasionally break, feed, shoe, and physic our own horses, and are daily brought into much more intimate and responsible relations (if I may use the term) with the animal than are any of the several persons who busy themselves with him in England. The groom, the farrier, the jockey, &c., in England, are probably unsurpassed, each in his line, but there is no class in that country who shine, except in this limited way. And then, what Englishman in England needs a saddle-horse? How many are there in England who ride, even once in their lives, what may be called a journey? We notice in this country that the amount of work which a newly-arrived emigrant, who has been used to horses, and an experienced bushman will extract from similar horses is very different: very few days' work serves to pump out the horse under the guidance of the former, whilst the latter will leave him behind, and continue on for weeks after, if necessary. Did any one ever see a new chum starting on his first journey who did not think his knowledge of horses quite a perfection in its way, and who did not come to grief with his quadruped somehow before the week was out?

The journey-horse, indeed, seems never

to have been much used in England. No doubt, as we learn, our ancestors were accustomed, on rare occasions, to ride their horses (having first made their wills) 100 or 150 miles from home and back again, at the rate of 20 or 30 miles a day. These, at best but short trips, were, however, of too infrequent occurrence to produce a ripe knowledge on the subject.

In offering the reader my ideas, I have been careful to confine my descriptions to some of those breeds of horses which are particularly worth an inquiry, and of which I have had generally some personal experience in their own country. Where such has not been the case, I have brought forward the opinions of authors who were themselves eye witnesses, and probably competent to form a correct estimate of the animal they describe: for I admire the precaution of the author of "Les Natchez," when he says, "*J'ai été voir les sites que je voulois peindre.*"

With regard to the possession of such experience as might be supposed to give the author some fitness to treat of the subject which he has undertaken, it may be allowable to state, that, a native of the colonies, circumstances have led me during twenty years to pass more time in the saddle than falls to the lot of most men : that the journeys I have made on single horses, extending from 100 to 1,000 miles each, are innumerable—and that I have lived amongst people of similar occupations, and have had the benefit of their experience. To this I have added some personal experience of the horses of England, Ireland, France, Belgium, Spain, Italy, Switzerland, Turkey, Syria, Palestine, Greece, Egypt, the Ionian Islands, the Cape de Verd Islands, Brazil, and New Zealand, as well as those of Tasmania and Australia. Besides the varieties of treatment and performance which these opportunities have placed before my eyes, I have witnessed the performance and sufferings of the horse under almost every phase of labour, accident, and hardship, with the exception of the battle-field. With the object I have had in view in writing these pages, viz., to point out the necessity for the production of a breed of *pure saddle-horses* in this country, as well as the means to that end, my treatise has naturally divided itself into three parts. In the first, I have endeavoured to place before my readers an unbiassed account of the English, Arabian, Andalusian, and Australian horses, together with a consideration of some of our best writers on the horse. In the second part, it has been my object to place, in a logical point of view, the true results of those axioms which have been received on all hands. And in the third part, I have endeavoured to point out how these axioms can be applied correctly and usefully to the circumstances of our country.

Since engaging myself with this subject, and consulting various works upon it, it has not failed to surprise me that I have found the customs and maxims of the East much more in accordance with the results of my own experience than those of the English, which have constantly led me to back my assertions by those of Eastern authorities. Had racers, instead of saddle-horses, been my theme, it would of course have been otherwise.

In directing the attention of my reader to Stonehange, my subject has led me to those portions of his work which I think are its weak points. His forte is the Racer, which animal, as is evident, notwithstanding his assertion to the contrary, is never absent from his view while writing, *and that not as a means, but as the great end of horse-breeding.* In both of these points I have quite disagreed with him. His Tenth Chapter is no doubt the pearl of his work, which, though the first of its class, is still in great measure *a réchauffé.*

Les Chevaux du Sahara, by General Daumas, is another work which I have considered at some length. It is more new, suggestive, and sound, as concerns the saddle-horse, than any I have met with in our own language, and as such, I recommend its perusal to my Australian reader.

I am afraid lest the latter half of my

book be considered egotistical by my readers. The perpetual recurrence of "I believe," "I have no doubt," &c., were, however, almost unavoidable expressions, when the opinions advanced were merely personal, and had no claim to be put forth as accepted facts. These, however, with my many other shortcomings, I must leave to the mercy of the reader.

Though what I have written is, I believe, correct (and is certainly the result of conviction), I am far from expecting that such will generally be acknowledged, or that it will be followed by any perceptible results. Some day, possibly, there will be a change wrought in what is now so imperfect. In the meantime, the ball has been set rolling by Lord Redesdale in England, the author has had the honour of giving it a push in Australia, and passing it on : it may get larger as it rolls—perhaps it may one day become an avalanche, and overwhelming our present rotten system, make room for the adoption of something more useful in its stead.

Let me here in conclusion give my thanks to those gentlemen who, by their subscriptions, have shared with me the expense of this publication.

MELBOURNE, 1863.

CONTENTS.

	PAGE
Preface	v

PART I.

Introduction	3
Youatt	8
Stonehenge, and the Saddle-horses of England	10
Cavalry Horses	54
The Arab Horse, and the Horses of the Sahara	71
The Arab Horse in India, Syria, Mesopotamia, and Persia	109
The Andalusian Horse	131
The Australian Saddle-horse	141

PART II.

On Blood	199
Selection of Sires	214
On Climate	219
Food	231
Recapitulation	239

PART III.

Saddle-horse Breeding in Australia	245
On Racing	266
On Riding Long Distances	289

PART I.

Tu es mon ami ; je te donnerais mes enfants : mais fais attention que mon cheval *c'est mon cou ;* si tu viens à me le ruiner, qui sauvera mes chameaux et ma famille au jour du danger.

Les Chevaux du Sahara.

INTRODUCTION.

> You do yet taste
> Some subtilties o' the isle, that will not let you
> Believe things certain.
> *The Tempest.*

Let us begin with the consideration of the English horse. The saddle-horses of England enjoy a great European fame. I will not ask whether they deserve their renown, but whether the perfection of which the saddle-horse is capable, has generally been realized in England? The question will, perhaps, raise the smile which comes of confidence and self-satisfaction with my English readers. I trust before we part to shake this confidence.

The following pages treat of saddle-horses, by which are meant roadsters and cavalry chargers; and also of saddle-horse breeding in Australia. As my convictions have forced me to disagree in important matters with received authorities on the first part of this subject, I shall endeavour to put the reader into a position to judge for himself of the correctness of my views, by placing before him some of the principal facts and observations which have led to, or rather confirmed, in my mind the conclusions which I shall advance. To this end, as something more than mere assertion (too much in vogue on this subject) is desirable and necessary, it will not be out of place to bring forward a succinct account of the condition, capabilities, and breeding of the English, Arab, and Andalusian horses, they being those which are held amongst a few others to have reached the highest state of perfection, before entering on the consideration of those of our own new country. From this examination it may be also possible, as a result, to fix with some degree of accuracy what the capabilities of a good saddle horse should really be, and to determine, by making past experience the basis of future efforts, what may be attempted by the horse-breeder, and on what principles, with legitimate hopes of success. For it is my belief that on an examination sufficiently careful and extensive, certain tendencies and results of a very important nature may be detected, and which hitherto may practically be said to have escaped observation, from which I cherish the hope that many time-honored prejudices may be exploded, and something more real, correct, and useful be substituted in their stead, and a solid foundation be thus laid for improvement. In support of such belief the production of facts will be doubly necessary, for it seems, indeed, hard to believe that anything new to an Englishman, or worth hearing, can be said of an animal that has been so carefully cultivated and described time out of mind. For in England and amongst Englishmen, a passing notice of the national favourite has not been thought beneath the dignity of either the historian or the legislator. He is placed before our eyes, on our coins, in our statuary, and our paintings. He is sung by our poets and exhibited in our theatres. For several centuries he has engaged the attention of devotees in every grade of life, from the throne downwards. Peers whose names arise as landmarks in our history, grave statesmen whose lungs one would think could breathe no other atmosphere than that of the House of Commons, have not esteemed it a small honor to be reckoned amongst the breeders of the best horses in Merrie England. Strains of celebrated blood have become heir-looms in families, and like other heir-looms have been accounted part of the glories of noble houses, until at length the English thoroughbred is believed by Englishmen to be the sanspareil of his kind.

There are, however, many in England and high authorities on the subject, who unwillingly have begun to chronicle the decline of the British horse, and denounce as ruinous the system under which he has been bred. This fact, which I shall endeavour to make clear, may justify, and perhaps lend some interest to inquiries on an animal that counts amongst us so many enthusiastic admirers.

Of all animals, the horse, without doubt, has always held the first place in the esteem of man. In America he did not exist at the period of its European discovery, and hence we miss his figure in the ruins of Palenque and Copan, but in all the other principal records, ruins, and remains of antiquity, he finds a place. At this day he is beloved by many in the palace, the cottage, the wigwam, and the tent. The Bible paints him in his glory, and he is not forgotten in the Koran. Still, as ever, he is the great auxiliary of states, both in peace and war. Can the consideration of him then in our new country be devoid of profit? Wisdom is not necessarily the dower either of old people or of old countries. Why then should we not judge for ourselves on the subject?

To have a prejudice disturbed, to be asked to reconsider a matter long set at rest; above all, to be asked to admit a whole string of errors, is, no doubt, a serious affair. These considerations, however, will not prevent me from endeavouring to show, that many of our ideas on horse-breeding are puerile and incorrect; that our present system is not the less wrong because based on errors which time and precedent have made respectable amongst us, or from trying to show that from the very nature of things, it is practicable to breed better saddle-horses on the Australian continent, than ever were, or can be bred in the green fields of Old England.

YOUATT.

> He is dead and gone. *Hamlet.*

Though Youatt, as an authority on the horse may be said to have made his bow to his countrymen and given place to Stonehenge, still, as his is the work on this subject most commonly found in the hands of Australians, I have thought it useful to say a word on his treatise. To Australians, inhabiting a continent two-thirds of the size of Europe, which on the whole bears much more resemblance to Arabia than to England, it becomes particularly pertinent to inquire, how we are likely to be served by Youatt's teaching, or our English preconceptions, as they affect the subject of horse-breeding, and to bear in mind, that were all that Englishman have written about their own horses correct, that still their knowledge and experience might require much modification when applied to our circumstances.

Youatt's treatise on the horse, useful in its day, though incomplete, has still served as the model and ground-work of the volume which has taken its place, and may claim paternity to many of the misstatements contained therein. As it may now be entirely said to have been superseded by Stonehenge, it will not require further notice at my hands.

STONEHENGE,

AND THE

SADDLE HORSES OF ENGLAND.

Our present breed of horses is undoubtedly less healthy than that of our ancestors; and this tendency to unsoundness is not marked in any particular department of the animal economy, but the defect shows itself wherever the strain is the greatest, from the nature of the work which the animal has to perform.

Stonehenge, p. 144.

The most recent, popular, and best work now in the hands of Englishmen on the horse is, "*The Horse in the Stable and the Field,*" by Stonehenge, and this author I have taken as the great authority on the horse of England, a position which cannot, I think, be accorded to him in connection with any other horse, except that of North America. As a successor to "*The Horse,*" by Youatt, his book is no doubt a great improvement. It contains more information, and takes more distinct and enlarged views, of part of the subject at least, rests on a more enlarged basis of facts, and analyzes more carefully received opinions. Some of the old pet errors of the English horse-loving public are also examined and clearly exposed, and future difficulties with present deficiencies in horse-breeding looked in the face with some fairness. At the same time the author who takes so large a subject as " *The Horse in the Stable and in the Field*," by which, as is clear from the context, the whole genus is meant, should, I think, have rested his assertions on a wider basis of information than is possessed by Stonehenge, if he hoped to treat the subject at all fully or correctly, and shown feelings more cosmopolitan and less decidedly English. So desirous, however, is this writer of showing the universality of his theme, that besides (as is usual on every subject now-a-days) giving the biblical history of the animal and drawing pretty freely on his classic lore, he even is so good as to furnish us with his synonyms in twenty different languages, which if neither very necessary to his subject, nor a very Mezzofantian exploit, is a beginning sufficiently ambitious to warrant the reader in expecting a great deal more than the writer either does or can furnish. But as my object is not to write a critique on Stonehenge, but merely to direct attention to the substance of some parts of his work, I will pass on, premising that that part of his book which treats of the veterinary art I have not read.

The first point with which I was struck on opening Stonehenge's work was the disparity of space allotted to the several breeds of horses. Thus it may be remarked that whilst the horses of England and what concerns them occupies 75 pages, irrespective of the chapters on Breeding, on the Brood Mare and her Foal, which refer almost entirely to England, as well as that on Stable and Stable Management, which bring the space allotted to the English horse alone and what concerns him up to about 180 pages, that the whole of what the author has to say of the horses of Asia, Africa, and Australia, is easily compressed into ten pages. Now, if the object of the writer was to place before his English reader that which the latter neither knows nor appreciates, rather than what many are already very well acquainted with, an arrangement the very reverse of this would perhaps have been more proper. To the horses of continental Europe are allotted four pages, whilst those of the United States of North America occupy twelve pages, and whilst the steeds of Circassia, Greece, Sicily, Brazil, China, and other countries too numerous to name are passed over in silence, the Brougham, the Gig'ster, the Phaeton horse, the Park Hack, and the Cob, receive a very full notice. Instead of all this biblical, classic, and philological lore, would it not have been quite as well if the author had simply told us that "the horse probably came out of the Ark in the days of Noah, but of his subsequent history I know nothing, until about that period of history when one Oliver Cromwell, familiarly called Old Noll, sat on the half-hidden throne of England, from which period up to date you will find me well informed in all concerning one class of Horses, viz., the Race-horse of England and of North America, on which subject I am able and willing to give you much information." Had he undertaken this task he could have acquitted himself of it well. In fact, this is what he may be really said to have done, and no more. The Racer is his hobby; to him, as a beau ideal, all others are compared; he usurps almost the whole of the author's attention, and is never absent from his mind when writing. As an authority touching racers, or the best strains of their blood in England and America, on the effects of in-and-in breeding, on crossing, on the *hit*, on the analysis of the pedigree of horses, I believe Stonehenge has no equal in our language, or in any other. These are his strong points; on all this he is clear, logical, and acute, and, except this, I know nothing worthy of note in his book. On other matters he has the average amount of information, and no more.

But whilst his information touching the racer is

so minute, great facts, and ones of much wider interest, seem to have escaped his observation; thus it is with surprise I find that, in treating of the Arab and other Eastern horses, he has overlooked the very important circumstance that there exist in the world, differing only in various places in the assiduity with which they are carried out, two distinct systems of horse breeding and management, we may call them the Eastern and European methods. That there exists indeed any system besides that in vogue in Europe, seems to have escaped the notice of our author, or at most to have only been received by him as one of those vague, disjointed, and incomplete rumours that require amplification and confirmation before they can be recognized as facts. Thus coasting, as it were, the confines of the subject, he writes:—" It is said that the Arab horse is only fed twice a day, but I conclude that this only refers to his allowance of corn, and that in the intervals he is allowed to pick up what little dry herbage the soil affords. . . . . About five or six pounds of barley or *beans*, or a mixture of the two, constitute the daily allowance of corn, which is the weight of about half a peck of good oats, and would be considered poor feed by our English horses, unless the proportion of beans is very large." All this, it appears, is *said* to be the case, and who says it? And where does the author learn it, or why does he surmise that *beans* form any or the principal portion of this corn? And why is not the reader entitled to a full and distinct understanding on this very important part of horse management? And why was not a system so different from ours, and so extensively practised as that in use amongst the Arabs, and in the case of so famous a breed, placed clearly before the reader of the *Horse in the Stable and the Field?* But if the horses of many large countries are overlooked, the hack and cavalry horses, certainly the most useful of saddle-horses of England, likewise receive but scant attention at the hands of our author; who, devoting himself, as I have said, almost entirely to the racer, awards, however, *en passant*, to every variety of English horse a superiority, which, except in the case of the thorough-bred, he does not attempt to substantiate. Let us, however, turn to some of the views and assertions of this author. "Grass," says Stonehenge, page 220, "is undoubtedly the natural food of the horse, though in his native plains the same species of plants are not to be met with as form the green surface of our own fields. English horses, however, may now be said to be thoroughly accustomed to our grasses, which seem to agree with these animals so well as to be one main cause of their superiority.

"A sound and moderately young animal gets fat during the summer and autumn months, when turned out on a good upland meadow; but he is not able to undergo long-continued exertion, especially at a fast pace; partly because the amount of fat accumulated in his internal organs interferes with his wind, but chiefly from the fact, that grass does not supply sufficient muscle-making materials for the wear and tear of his frame.

"Winter grass, which contains no clover, from this plant not being of a nature sufficiently hardy to stand the frost, is so void of nourishment, that the horse confined to it alone speedily becomes very poor, and will almost starve if he has not some hay and corn."

Let us take the above passage as a sample of Stonehenge's loose way of making assertions, and point out a few only of the important inaccuracies and doubtful positions which he here assumes as facts. "Grass is undoubtedly the natural food of the horse in his native plains, &c." Now, how does he learn that the horse is a native of the plain rather than the hills? Both from reading and observation I object to this assertion as incorrect, taking his present habits, when quite unfettered, as a proof of his present, and therefore original predilections. At page 1, indeed, he goes more into particulars, and assigns the plains of Central Africa as his original habitat. Without, however, favouring us with the source of his information, adding, "where curiously enough, he is not now to be found in his wild state." To this he might have added, where grass is scarce, the climate the reverse of that of England, and water not plentiful, therefore, very unlike the country where he is now said to attain perfection, and where, consequently, if this is correct, he must have been first placed in an ill-selected locality. But as I have just said, where there is no direct proof on the subject, why should we suppose him to have first inhabited plain country, where he must of necessity be inferior, rather than the hills, where alone experience teaches he ever reaches the full perfection of which he is capable? He continues, "English horses, however, may now be said to be thoroughly accustomed to our grasses, which seem to agree with these animals so well as to be the main cause of their superiority," and yet in the same breath he tells us, "that in winter the horse confined to grass alone speedily becomes very poor, and will almost starve if he has not some hay and corn." So that the grass to which the English horse owes his superiority is of such a sort, that he becomes too fat and wanting in muscle to work on it in summer and autumn, and is almost reduced to starvation on it in winter. Now in other countries, I hope to show, that on grass alone the horse attains his full size and strength, enjoys excellent health, and is much more capable of hard work than the horse of

England, (feed him as you like,) whilst entirely grass-fed. If, then, English grass is insufficient for the horse, English hay must naturally be so likewise ; and so Stonehenge informs us, that " between a race-horse reared on corn, and another confined to hay and grass, the difference in value would be a thousand per cent.;" further, that all horses in England require it, but that to give it to hacks, &c., is impossible on account of the expense, which he says would never be reimbursed. Practically, then, he admits, few horses can be brought to perfection in England, because the grass of that country is insufficient for the purpose, and the necessary artificial food too costly. This is, let me remark, a difficulty in the way of reaching perfection in horse-breeding, which is not universal, for it must be admitted that the grass on which the horse is always well, and on which he can work hard at all times, is more likely fully to develope his powers than the grass on which he cannot work in summer and autumn, and on which he nearly starves in winter.

At page 54 of the same author, we learn that in England, "gradually a breed of horses has been established, which has been celebrated through the world for the last century, for speed, stoutness, and beauty ; in all which qualities the present stock exceeds their parents on both sides. Much of this excellence is doubtless due to the climate and soil, which encourages the growth of those fine grasses, that exactly suit the delicate stomach of this animal." With regard to the delicate stomach of the horse, though, no doubt, it has been the fashion to speak in this manner in England of the whole genus, where indeed the horse, and especially the thorough-bred, has a very delicate stomach, yet I have never been able to discover any symptoms of delicacy about him in other countries, where his readiness to feed, his power of abstinence, and the variety of plants he is in the habit of feeding on, is quite as great as those of most other herbiverous animals. And yet, at page 19, he attributes much of the excellence of the horse of Arabia, to its sandy soil, light grasses, and dry climate. To me it appears, that if the cold, moist climate of England, together with its luxuriant grasses is a *disideratum* for horses, that then the dry atmosphere of Arabia with its poor grasses and arid sands should be their *bane* ; or that grass and climate are circumstances of little import to them.

But if Stonehenge makes rash assertions as here in the matter of climate, he generally, sooner or later, gives you the truth of the matter. So in this instance, " Shelter from the weather," says he, " should, however, be afforded to colts of all classes during the winter season, and, unless they have this they soon grow out of form, however well they may be fed. A colt neglected in its first winter never recovers its proper shape, nor does it grow into the size and strength of body and limbs which naturally appertain to its breed." Now, I think that a climate which is so severe as to require the horses bred in it to be housed, to prevent them growing out of form, cannot be said to be particularly adapted to that animal. Hence, if the English horse is superior to all others, he can manifestly owe it, neither to grass without corn, nor climate without shelter.

His speed I have allowed; as for the claim to stoutness and beauty we will examine that in a moment; in the mean time let me ask in what does the superiority which we so often hear claimed for the English horse consist? It cannot be on the score of racing—in which he is without a rival—because the speed of the race-horse serves no useful purpose of life, is too costly in its attainment, is debarred to all but boys and dwarfs, and indeed is a means, not an end. To this we shall see in a moment that Stonehenge agrees. Is it then on his appearance or beauty? but of what use is his appearance if it only belies his performance? If he looks a good horse, and at work is found to be a bad one? "The great object," says Stonehenge, " of encouraging the breed of racehorses is" (not to race) "however lost sight of, if suitable crosses for hunting, cavalry, and hack mares, cannot be obtained from these ranks. In these three kinds, soundness of feet and legs is all-important, together with a capacity to bear a continuation of severe work." Hence we must conclude that a racer is not a hack; second, that a racer not fitted to get hunters, hacks, or chargers, is a failure. Then, let me ask, what breeder of thoroughbreds hopes that his mare will throw a colt fit to be a sire of saddle-horses? we all know he flies at higher game, if not so useful. His interest is to breed a winner of races; the sire of saddle-horses is not his hope, but his misfortune. The racer pays him better and brings more renown. And thus in fact we learn from Stonehenge, not that the object of breeding racers is lost sight of, if these racers are not fitted to be sires of saddle-horses, but that "the great object of encouraging the breed of race-horses is, however, lost sight of, if suitable crosses for hunting, cavalry, and hack mares *cannot be obtained from their ranks.*" Hence the occurrence of such stallions is not the rule but an accident; so that racing has a happy accident for its great object. We shall see how rare the accident is.

Now, if there is one point constantly and strongly insisted on by Stonehenge, and well known to every breeder, it is "*that like produces like, or the likeness of some ancestor.*

That the foal is in a great measure indebted to the sire for his qualities, good and bad, is a truism found in the mouths of all horse breeders of every country. Hence the careful selection of sires, for if such were not the case one sire would do as well as another, and all breeding be a lottery. It is further known to all and received by Stonehenge, that not only are speed, stoutness, soundness, &c., and their opposites inherited chiefly from the sire, but that even acquired qualities are in part inherited from him, which indeed must be apparent to all who have experience in animals as well as birds of any description. Then if "like begets like," on what principle is the Racer used as the sire of Hunters, Hacks, or Chargers, in which horses qualities very different from his own are necessary? In truth, this is a direct contradiction of the great principle on which is founded the whole theory of breeding. Let us go a little into particulars. In hacks, hunters, and chargers, we are told (and who can deny it?) that "soundness of feet and legs are all important, together with a capacity to bear a continuation of severe work." In other words these horses must be sound and stout. To be so, they must—like begets like, remember—be got by horses that are sound and stout; they are, we have seen, got by the thorough-bred or racer. Well! is he sound and stout? Whatever he is, they must be to a considerable extent, and as such you must receive them, or cast aside the received theory and substitute another on the subject, different from that received either in England or anywhere else. You will remind me of the dam and her influence on her progeny. Admitted. But as the one cause only can be at work with both parents, it simplifies the inquiry to consider the sire alone, as his peculiarities, good or bad, must in a great measure, at least, be transmitted to the foal. Then I repeat, what is true of the sire must, as a rule, apply to his get. Has then the racer sound feet and legs? Is he sound and stout? Listen to Stonehenge, speaking of English horses (page 81). " But how many of the fashionable sort," says he, " will bear constant use on the road without becoming lame? and how many sound horses are there to be met with out of a hundred, taken from the ranks of any kind *tolerably* well bred? Every horse proprietor will tell you scarcely five per cent., and some will even go so far as *to say that a sound horse is utterly unknown.*" The same author (page 80) further says, " One chief difficulty of the trainer now is to keep his horse sound, and unfortunately as disease *is in most cases hereditary,* and too many unsound stallions are bred from, the difficulty is yearly on the increase. Without doubt, roaring is far more common than it used to be, and the possession of enlarged joints, and back sinews, is the rule instead of the exception. During the last ten years, the Derby has five times been won by an unsound animal, which the trainer was almost immediately afterwards obliged to put out of work, either from diseased feet or a break-down, and yet few breeders think of refusing to use such horses as these."

The following chapter quoted entire from Stonehenge, page 144, will interest the reader.

"IMPORTANCE OF HEALTH AND SOUNDNESS IN BOTH SIRE AND DAM.

"Our present breed of horses is undoubtedly less healthy than that of our ancestors, and this tendency to unsoundness is not marked in any particular department of the animal economy, but the defect shows itself wherever the strain is the greatest, from the nature of the work which the animal has to perform. Thus, the race-horse becomes a roarer, or his legs and feet give way. The hunter fails chiefly in his wind or his hocks, because he is not used much on hard ground, and, therefore, his fore legs are not severely tried, as in the case of the racer, who often has to extend himself over a course rendered almost as hard as a turnpike road by the heat of a July or August sun. The harness-horse often becomes a roarer, from the heavy weights that he has to draw, especially if his wind-pipe is impeded by his head being confined by the bearing-rein. The hack again suffers chiefly in his legs, from our hard macadamized roads, whilst the cart-horse becomes unsound in his hocks or his feet, the former parts being strained by his severe pulls, and the latter being battered and bruised against the ground from having to bear the enormous weight of his carcase. But it is among our well-bred horses that unsoundness is the most frequent; and in them, I believe, it may be traced to the constant breeding from sires and dams which have been thrown out of training in consequence of a break down, or 'making a noise,' or from some other form of disease. It is quite true that roaring is not necessarily transmitted from father to son, and it is also manifest that there are several causes which produce it, some of which are purely accidental, and are not likely to be handed down to the next generation. The same remarks apply to the eyes, *but in the main it may be concluded that disease is hereditary*, and that a sound horse is far more likely to get healthy stock than an unsound one. In the mare, probably, health is still more essential, but if the breeder regards his future success, whether he is establishing a stud of race-horses, or of those devoted to any kind of slower work, he will carefully eschew every kind of unsoundness, and especially those which are of a constitutional character. If a horse goes blind in an attack of influenza, or if, without any previous indications of inflammation, he breaks down from an accidental cause, the defect may be passed over, perhaps; but, on the contrary, when the blindness comes on in the form of ordinary cataract, or the break-down is only the final giving-way of a leg which has been long amiss, I should strongly advise an avoidance of the horse which has displayed either the one or the other. I believe that a Government inspection of all horses and mares used for breeding purposes would be a great national good, and I look forward to its establishment, at no distant time, as the only probable means of insuring greater soundness in our breeds of horses. I would not have the liberty of the subject interfered with. Let every man breed what he likes, but I would not let him foist the produce on the public as sound, when they are almost sure to go amiss as soon as they are worked. Ships must now all be registered at Lloyds in the classes to which they are entitled by their condition; and horses, as well as mares, should be registered in the same way, according to the opinion which the Government inspector may form as to their health, and the probability of getting or producing sound and useful foals. The purchaser would call for the registration mark, when he asked for the pedigree of the horse he was about to buy, and if it was not a favourable one, he would, of course, be placed upon his guard. If this plan could be carried out in practice, as well as it looks on paper, much good might be done, I am assured; but we all know that inspectors are but mortals, and that they are liable to be biassed in more ways than one; still, I believe that the evil is becoming so glaring that something must soon be done, and I see no other mode so likely as this to be advantageous to the interests of the purchaser and user of the horse." What will an admirer of English horses think of all this?

In training, the superfluous fat used to be removed from racers by sweating gallops with cloths on. For the gallop the Turkish bath has now been substituted, with the object of sparing the animal's legs. "No wonder, therefore," says Stonehenge, page 264, "that trainers eagerly resort to the use of the bath, as every year their horses seem to be getting more liable to break down." But this is not all, nor the worst; so delicate and frail is the English high-bred horse, that we find, that not only is he unable to sustain his sweating gallops and his races uninjured, as we have seen, but he can no longer even be trusted loose on his summer pasture ground. "Not only," says Stonehenge, page 276, "is his stomach pinched, but his legs and feet are damaged by being battered on the dry soil.

The feet of wild asses, and even native Arab horses, may be able to bear the blows and friction of the wastes over which they travel, but those of English horses are undoubtedly not formed of such strong and tough materials, so that it is utterly unwise to leave them exposed to the risk." To this evil he again proposes the interference of Government as a remedy. "We are sadly in want of sound, well-bred stallions for general purposes, and if the Government of the country does not soon interfere, and adopt some means of furnishing these islands with them, we shall be beaten on our own ground, and shall have to import sound, useful horses from Belgium, France, Hungary, or Prussia, whichever country can best afford them."

Whatever then may be the attributes of the English horse, whose superiority is mainly attributable to the grasses and climates of England, soundness, or the first requisite, is not one of them. The other vital necessity in a good saddle-horse is "*a capacity to bear a continuation of severe work*," in other words *stoutness*. Before, as I may say, taking evidence on this point, I would ask, can an unsound horse be stout? Will not a continuance of hard work be sure to bring a break-down? It is true that a horse may be quite deficient in stoutness and yet sound ; but he cannot, it appears to me, if all his limbs are necessary to him, and there exists no superfluity in his conformation, be unsound and yet stout. In the instance before us, we shall see the cause followed by its effect—the tree loaded with its fruit. "I wish now," says Stonehenge, page 81, "to impress upon my readers that while the race-horse of 1860, is as fast as ever, as *stout* as ever, and as good looking as ever, he is made of more perishable material, in proportion as he comes to maturity at an earlier period. Any of our modern two-year-olds would probably give two stone and a beating to Eclipse at the same age, but if afterwards they were put to half-bred mares for the purpose of getting hacks, chargers, and hunters, the stock of Eclipse and Childers would be much more valuable than any which we have at present. We sadly want sound and well-bred stallions, &c." Hence the thoroughbred in his useful capacity, viz., that of a sire of saddle-horses has deteriorated. In this passage the sire of our saddle-horses is said to be as stout, as fast as ever, and yet more perishable than of yore. But, is he in fact as stout as ever, or is this assertion only a little ebullition of the patriotism or partiality of the writer? Stonehenge, as usual, supplies his own refutation, for this author is ruled in his writing by two clashing feelings; partiality to the horses of his country would make him hide their defects, whilst the hope that their exposure may bring a remedy, induces him to lay them bare.

He would throw a cloth over the imposthume, and strike the lancet through the covering. I am less particular, and seek only the truth. Let us hear what he says, "In spite, however (page 79), of all the elaborate calculations which others AS WELL AS MYSELF have made, I cannot quite divest myself of the belief that Lord Redesdale is correct in his assumption that the thoroughbred horse of the present day is, on the average, less stout than he was of yore. That there are some few which can race and also stay, I firmly believe, and that many which cannot race, but can stay, are early drafted into the hunting stable is also my opinion, *but that the majority are deficient in stoutness seems to me a patent fact.*"

Again, "I am quite of opinion (page 79), that taking any number of race-horses at random in the year 1860, they will not on the average bear comparison, in point of stoutness, with a similar number, either of the year 1800 or of the year 1760." I could adduce much more from this author to the same effect, but the above fully marks his belief, and so answers every purpose, for it leaves no escape from the conclusion, that the climate, grass, system of breeding in England, or some union of these causes, have produced an imperfect and daily deteriorating horse, unless the reader leans to the opinion that a delicate and unsound animal is preferable to a healthy and sound one. Let the reader quietly re-consider the various extracts which I have made from Stonehenge, and he will, I feel convinced, if he accepts that author as an authority, see the fairness and truth of my deductions and conclusion. Many travellers who have had opportunities of seeing the horses of other countries take an exactly similar view of the subject. One or two instances of this, as Captain Nolan and Captain Shakspeare, are adduced in the sequel, *whereas I have never met with an author, who, throwing aside mere assertion, undertakes to show by comparison with other horses that the saddle-horse of England is good or stout,* and I believe that there is no instance on record in which it is attempted to be shown by a credible eye-witness that a body (for I have nothing to do with rare exceptions or picked specimens) of English horses have earned for themselves the title of stout saddle-horses. On the contrary, whatever evidence I have met with on the subject has been all the other way; so that if the great principle of horse-breeding that "like begets like" is admitted, and the evidence I have brought forward proves that the racer who is the sire of the English saddle-horse is neither sound nor stout, then it must follow that the English saddle-horse is unsound and soft. And, if so, in what consists his *superiority*, of which it is the fashion to talk so much? There is one circumstance which I think alone causes this to be doubted by many Englishmen, that they have never been in a position to compare their own with other horses, and that they do not know what a really good horse can do or suffer; but are like persons who have lived in an hospital till groans seem music, and jalaps and squills the proper drinks of healthy men.

But, besides, when actually at work we find the saddle-horse of Great Britain displays in various ways the frailty and debility of his constitution. Thus Stonehenge, for instance, when speaking of the small stomach of the horse, page 235, says:— "From the first of these causes the horse must never be allowed to fast for any long period if it can possibly be avoided, it being found from experience, that at the end of four hours his stomach is empty, and the whole frame becomes exhausted, while the appetite is frequently so impaired, if he is kept fasting for a longer period, that when food is presented to him it will not be taken." This may be true in England—no doubt it is true there, more especially of the racer, the true pole to which the thoughts and remarks of the author constantly turn, and in favour of which all other horses are constantly forgotten, but if it be meant to apply to the horse universally, then the Arab horse must be considered to pass his whole existence in a very exhausted state, for he never breaks his fast but twice in the 24 hours, and even then in a manner which an Englishman would consider but sparing, as the whole operation does not occupy him two hours daily. "*Il est rare qu' on donne à manger le matin,*" says Abd-el-Kader, "*le cheval marche avec la nourriture de la veille, et non avec celle du jour.*" The same may be said of the Eastern horse in general.

Neither does this precautionary injunction enter much into the calculations of Australians, as we ride our grass-fed horses thirty miles a day for twenty successive days without any difficulty or any such precaution.

We have seen, from Stonehenge, that the great object sought in the cultivation of the pure thoroughbred horse of England, when not lost sight of, is to produce a horse eminent for all those qualities which are desiderated in a saddle-horse, and which it is expected he will transmit to his offspring; but the first anomaly that strikes me in all this is that the saddle-horse himself has no pedigree; his escutcheon is always crossed with the bar sinister. To me it appears unreasonable that if pedigree is all important in the sire of the saddle-horse, and that "like begets like," that it should be neglected in the beast actually to be ridden. Seeing, however, that the thoroughbred is unequalled in speed for a short race, carrying a feather, and is the sire of the cavalry horse, hack, and hunter of England, and that he seems destined to the same end in France, Germany, Russia, and America, it becomes interesting and important to breeders to scrutinize his pedigree, and to ascertain by what means he has attained that unrivalled speed which he undoubtedly possesses. In matters of fact on this head no one equals our old friend Stonehenge, but as he is not always quite straight forward, let us, as usual, cross-question him a little. He says, page 82 :—" In the first place purity of blood must be considered as a *sine qua non*, for without it a horse cannot be thoroughbred, and therefore we have only to ascertain the exact meaning of the term 'blood.' It is not to be supposed that there is any real difference between the blood of the thorough-bred horse and that of the half-bred animal, no one could discriminate between the two by any known means; the term blood is here synonymous with *breed*, and by purity of blood is meant purity in the breeding of the individual under consideration; that is to say, that the horse is entirely bred from one source, is pure from any mixture with any other, and may be a pure Suffolk Punch, or a pure Clydesdale, or a pure thorough-bred horse ? (*and why not a pure saddle-horse ?*) But all these terms are comparative, since there is no such animal as a perfectly pure-bred horse of any breed, whether cart-horse, hack, or race-horse; all have been produced from an admixture with other kinds, and though now (how in the matter of the saddle-horse I should like to know?) kept as pure as possible, yet they were originally compounded from varying elements; and thus the race-horse of 1700 was obtained from a mixture of Turks, Arabs, and Barbs. Even the best and purest thoroughbreds are STAINED (remark the word for it is of consequence) with some slight cross with the old English or Spanish horse, as I have shown at page 54; and, therefore, it is only by comparison that the word pure is applicable to them or any others. But since the thoroughbred horse, as he is called, has long been bred for the race course, and selections have been made *with that view alone*, it is reasonable to suppose that this breed is the best for that purpose, and that a stain of any other is a deviation from the clearest stream into one muddy, and therefore impure; the consequence is, that the animal bred from the impure source fails in some of the essential characteristics of the pure breed, and is so far useless for this particular object. Now, in practice, this is found to be the case, for in every instance it has resulted, that the horse bred with the slightest deviation from the sources indicated by the stud book, is unable to compete in lasting power with those which are entirely of pure blood. Hence it is established as a rule, that for racing purposes every horse must be a thoroughbred, that is, as I have already explained, descended from a sire and dam whose names are met with in the stud book." On the same subject we read, page 71, " To define the thoroughbred horse of the nineteenth century is easy enough, because it is only necessary to adduce the law that he must appear in the stud book. Without this testamentary evidence no other will be received, nor even theoretically can any other be adduced. By some it is supposed that he is a horse descended from sires and dams of Eastern blood, that is, either Turks, Barbs, or Arabs; but this has long been known to be a fallacy, for we find numerous gaps in almost all the old pedigrees, which there is every reason to believe ought to be occupied with the names of native or Spanish mares." A few lines lower down, page 71, he adds:—"An Eastern horse is at once admitted (into the stud book) as being supposed to be of pure blood, and there is no difficulty in his case, &c." Now with these definitions and remarks I have several faults to find. The definitions of "blood," and of the thoroughbred, do not to my mind distinctly or even correctly dispose of the question. For example, let us place side by side the two following passages:—"To define the thoroughbred horse of the nineteenth century" page 71, "is easy enough, because it is only necessary to adduce the law that he must appear in the stud book. Without this testamentary (*documentary ?*) evidence no other will be received, nor even theoretically can any other be adduced." Now this is monstrous! If a half-bred were entered in the stud book, would the entry purify his blood? Would West Australian be less a thoroughbred, if his name had not been entered? The stud book is no doubt convenient. Entry in the stud book may constitute the only evidence which will be received of any horse being thoroughbred but certainly it is no definition of what that animal is. It is the difference between a fact and the proof of such fact. Besides it is only in the next page that the author says :—" There is no doubt that when half-bred races were in fashion, numerous exchanges of foals took place, by which the thorough-breds were made to appear as half-bred, and *vice versa*." Even as a record, then, the stud book is not immaculate, but still less can it be considered any definition of the thoroughbred, or any perfect record, when we read in the same page an Eastern horse is at once *admitted* as being *supposed* to be of pure blood. The public have agreed to receive the stud book as a definite authority on the matter and no more. But after all this want of correctness may practically be of small consequence. Not so what follows, viz. :—" By some it has been supposed that he (the thoroughbred) is a horse descended from sires and dams of Eastern blood, that is either Turks, Barbs, or Arabs, *but this has long been known to be a fallacy*, for we find numerous gaps in all the old pedigrees, which there is every reason to believe ought to be occupied with the names of native or Spanish mares." Now on the point concerning which the author has just emphatically told us that none but "testamentary evidence" will be received, how are we to be expected to receive an assertion so loosely made as this? What proof does the author give us that there is a fallacy in attributing an unmixed Eastern descent to the thoroughbred? It is at best but putting one supposition against another. Besides, does a breakdown in a pedigree necessarily lead to the inference that the gap should be filled up with the name of a native *or* a Spanish mare? Is it even probable that it should be filled up with the name of a native mare? And of what blood would be the Spanish mare? Could it be Eastern? These points I shall examine in a moment, though I should have preferred "testamentary evidence" to any other from Stonehenge—the "pound of flesh" from Antonio! But not only was a gap to be filled up, but it was filled by what "made the charm firm and good," according to our author, who adds, page 56, "Indeed, I believe that the use of the Spanish, mixed, *perhaps, with the native English blood* in the mare, was the real cause of the success which attended the cross with the Barb; the mare being of greater size and stride than the horse, and giving those qualities to the produce, while the horse brought out the original strain of Eastern blood, which possessed the wind and endurance so peculiar to it." So that the size and stride are said to be the English contribution to the racer, whilst wind and endurance are his heritage from the East. In this, however, and several other passages, Stonehenge evidently wishes to rebut the idea of the thoroughbred horse being wholly of Eastern blood; and further than this, he would leave the impression that it was by engrafting the Arab blood on the English stock that has made the *perfection* of that horse. Though he says, as I have shown a few pages back, that "even the best and purest of our thoroughbreds are *stained* with some slight cross of the old English or Spanish horse." Here is a jumble! What are we to believe? First, purity of blood is held up for our admiration, then a stain is said to improve a character! Of this junction of old English blood, however, he advances no proof. Indeed, when this special pleading is laid aside, and he is off his guard, he tells us, point blank, of the thoroughbred, page 155, "Being a native of a warm and dry climate, he requires to be protected from the weather, &c." I presume England is neither a warm nor very dry climate. The simple truth is, that certain pedigrees are incomplete; of what blood were the unrecorded animals remains a matter for conjecture. It has commonly been held to be

Eastern, Stonehenge wishes to have it thought English. Now all the blood that is recorded in the stud book is Eastern, all the known roots of that breed are Eastern, and as we know that Eastern blood had constantly been imported into England, and its value, though fluctuating, appreciated there from the days of the Norman conquest to the date when the English pedigrees commence, is it not extremely probable that the unknown blood was, in a great measure at least, Eastern also? And if the racer of this day is more fleet and larger than his desert ancestor, can it be attributed to nothing more reasonable or likely than an admixture of blood which has always been slow? I thought Stonehenge had told us, page 82, that "purity of blood must be considered as the *sine qua non*, for without it a horse cannot be considered thoroughbred?" But here he would make his *improved* horse a mongrel—improved, be it remarked, by, and not in spite of, his cross. But, worse than this, to admit that inferior blood can improve what is superior, overthrows the whole system he has so emphatically laid down; which leaves us only two alternatives—either that the thoroughbred is not superior to the Eastern horse, or that, if he is, the superiority cannot be in consequence of an amalgamation with base blood, but in spite of such amalgamation, and attributable to some other cause. Now, a superiority in speed is very easy accounted for, in conformity and not in contradiction to a well-known law, and one on which Stonehenge himself insists, without the necessity of having recourse to base blood, viz.: by a constant, and for that purpose judicious selection for sires, of those horses which have proved themselves eminent in this point. The English having done this, selected sires for speed alone for 150 years, and the Arabs, not having pursued this plan, requiring speed of quite a different sort, and united with qualities not sought on the turf, it is but a natural result, as any one conversant with animals will allow, that the one should have become more speedy than the other.

Nor is his increase in size more difficult to account for than his speed, without any recourse in theory to base blood. If it had come from a cross with a mare larger than the Arab, it would have been immediate, and probably the first cross larger than any other, as I have personally witnessed in breeding horses, sheep, and cattle. But this was not the case; the increase has been gradual and progressive, as we learn from Stonehenge himself, who says (page 72), "Out of 130 winners in the middle of the last century, only 18 were fifteen hands and upwards, whereas, now, a winner below that height is a very great rarity indeed, even among the mares. *The increase of size is, no doubt, mainly due to the influence of the Godolphin Barb, who was* himself larger than most of the *Eastern sires,* and got stock of a still greater height. His son Babraham was fifteen hands high, then considered an extraordinary development; of the eighteen winners mentioned above as being fifteen hands and upwards, eleven were by the Godolphin Barb or his son."

Thus as Babraham was not of English blood, he upsets completely his previous assertion concerning English blood, the increased size being attributable, not to English mares, but to the Godolphin Barb in the first instance, and having been very gradual, as he asserted, and as is well known. Still further causes and no doubt the principal ones are, the increased quantity of food which he has been allowed since his importation to England, and also from the *effects of a climate which gives to stature and bulk what it takes from solidity of muscle and density of bone.* For as Captain Shakspeare tells us, in his description of the Arab horse, as compared with that of England, " The bone looks small, but then it is very dense; the hollow which contains the marrow being very small and the material solid, more like ivory than bone, heavy and close grained." He also speaks of the muscles feeling firmer to the hand. I feel no doubt, indeed, that it would be found on experiment that the Arab is of greater specific gravity than the thoroughbred. But, after all, is the thoroughbred an

Eastern or a cock-tail ? Let us hear Stonehenge once more. "*My own belief*," says he, page 55, "is, that the race-horse of that (Eclipse) day was imported from Spain, and bred from a cross of the Andalusian mare with the Barb introduced by the Moors." Here, too, we have no mention of English blood. Since those days no other blood than Eastern, has knowingly at least, been added to the stud book, but with Eastern blood it has been so deluged, that any baser alloy did it ever exist, must long since have been drowned, which will not be difficult to credit, when it is shown by Stonehenge that the offspring of a common mare, may, in *twenty years*, be so far reclaimed as to leave only one sixty-fourth of the original alloy of the dam. Thus we trace the most esteemed English horses practically to Eastern blood alone, and it becomes neither unimportant nor without interest to examine how these foreign progenitors of our racers were themselves bred, a subject on which our author is silent. The first question naturally asked is, are the Barbs, Turks, and Arabs, of which England has imported so many, one race, or are they so many distinct races ? Allow me on this subject to offer the ideas of General Daumas, from "*Les Chevaux du Désert.*" "If we might be permitted to offer our personal opinion, (which he very rarely does,) we should say, that there is a disposition to draw too marked a line between the Arab and the Barb horse. A more comprehensive name, that of the Eastern horse, should, it appears to us, be given to them both; for they are of one great family of common origin, modifying and displacing each other under the influences of various climates, which after all are not very different.

"Strength, activity, vigour in conformation and in action, is the dower of the horse the moment you meet him on this side of the Euphrates, and on the other side of the Mediterranean and the Caucasus, where he remains on the soil of Islamism; there he is always muscular, abstemious, and invincible to privation and fatigue, living betwixt heaven and the sands. Call him if you choose, Numidian, Barb, Syrian, Arab, Persian, or Nedjd, it is of little moment, all these are but Christian names (*prénoms*), if I may be allowed the expression, his family name is the horse of the East. The other family, this side of the Mediterranean, is the European breed."

It is natural, in fact, that as the necessities of the riders of these various horses are very much the same, and the circumstance of climate and mode of treatment very similar, that there should exist a great family resemblance amongst their horses. And such is found to be the case, so that the remarks made on one variety, will, to a great extent apply to all. The Arabs, indeed, have no stud book, though records of very ancient date of some of the most esteemed families of their horses are well known to exist, but amongst a people so much wrapped up in their horses as they are, it is easy to conceive that their descent will be well known. Abd-el-Kader, in his celebrated letter to General Daumas, says, page 407, "You ask me if the Arabs of the Sahara have registers wherewith to establish the filiation of their horses." "Know then that the people of the Algerian Sahara do not trouble themselves with registers anymore than do those of the Tell. Notoriety is enough for them, as the genealogy of their pure horses is as well known to everybody as is their own." But, after all, amongst a poor people necessity is the great source of their customs; and so we find it with the Arabs. "The nature of the Sahara horses," says the Emir Abd-el-Kader, "is a consequence of the lives of their masters; scarcity of food obliges the Saharians to accustom their horses to hunger, and to thirst from lack of water, which is often not found nearer than one or two days' journey from their encampment. Their speed and endurance of fatigue, arises from the numberless quarrels of these Arabs, from their incessant war excursions, and their love of hunting the fleetest animals, the ostrich, the gazelle, and the wild ass, which some of them will continue for a whole year uninterruptedly." When, from any circumstance, as the capture of a horse in battle or otherwise, doubts arise as to the purity of a horse's blood, they are decided by trial as well as by his appearance. On which subject the Emir says, "Where there is no public notoriety on the subject, it is by a *proof of speed joined with bottom* that the Arabs judge of their horses, and of the nobility and the purity of their blood, but their figures also reveal their qualities." What is the amount of speed and bottom required from horses of pure blood will appear in subsequent chapters, and will, I have no doubt, be deemed satisfactory and to spare to the most fastidious. I indeed believe myself, without wishing to question the undoubted utility of a stud book, that performance, if not an irrefragable proof of purity of blood, is its best witness, and the only guarantee that *whilst pure it is not degenerate.* That under-bred horses would often be able to compete with thoroughbreds is not likely, for as Stonehenge notices, page 82, "in every instance it has resulted that the horse bred with the slightest deviation from the sources indicated by the stud book is unable to compete in lasting powers with those which are entirely of pure blood."

Our thoroughbreds, then, are descended from the Eastern horse, which was bred by equestrian nations, with infinite care and judgment, on the strictly applied principle "*that like begets like, or the likeness of some ancestor.*"

I have thought it right at this place to say this much about the Arab horse, and have only to add, that whether a first-class Eastern horse or mare has ever been brought to England is a matter of much doubt: so that it appears that the thoroughbred is descended from Eastern horses of what class or quality is unknown.

In quoting the opinion of the Emir Abd-el-Kader, not only on the Barb but also on the saddle-horse in general, I have accepted him as the very highest authority on that animal. Whether we look at his admitted judgment or skill on the subject, which is notorious and proverbial amongst Eastern nations and well known in France, or reflect on the great struggle he carried on for years through the agency of his horses, as well as the soundness of the principles which he has given us on the subject, we are bound, I think, to accept him as an authority second to no man. This much it has seemed to me necessary to say at present of the Arab horse, as connected with the English thoroughbred, and through him with the English and Australian saddle-horse. With regard to Stonehenge, it will not be needful to press further the consideration of his work. His enthusiastic admiration of the thoroughbred leads him into inconsistencies. He upholds him as the first of horses, and yet allows it to leak out that he is delicate, unsound, and useless. To his agency as a sire (added to "grass so admirably fitted to the delicate stomach of this animal" that it renders him too fat in summer and autumn and too poor to work the rest of the year, joined to a climate so excellently suited to him as to require to be hourly guarded against) he attributes that fine breed of horses, which he himself considers to have been long deteriorating, and which he in fact shows us will in the future require government support, with the help of sires from France, Belgium, Prussia, and Hungary, to preserve from even a fuller decrepitude. But even this partiality and folly might have been passed over, and must be esteemed but a venial offence compared to his great and radical contradiction—that of first singing the praises of pure blood in horses of *every* description, and yet holding up as the acmé of possible perfection as a saddle-horse the bastard offspring of the thoroughbred.

In two words, I can only estimate this author as an able exponent of an utterly untenable system.

In offering my concluding opinions and remarks on the saddle-horse of England, I will do so first absolutely and then as compared with the Eastern horse.

It will then be understood that he is absolutely a cross-bred animal, save in a few exceptional cases, where a "sticket" racer is drafted into the hunting stable, and hence, if there is any meaning in the words *pure blood*, and the stress laid on them by all writers, he is necessarily and inevitably inferior; inferior, if to none other that exists, as his friends would have us believe, at least to the possible of perfection He is got by the thoroughbred out of any mare, and is only accidentally a saddle-horse, for as far as his blood is concerned, he might have been hack, hunter, or charger, gigster, park-horse, or coacher; or have filled one of the other fifty refuges, which are open to his reception. It is a fact which will perhaps, some day, be acknowledged that whatever good is in him he owes to his Eastern blood, and, *in spite of the climate in which he is reared*, and in spite of the system under which he is bred, which is quite as much that like begets unlike, as that like begets like. Unfortunately for the country, a few exceptional horses, which seem to have preserved some of the stoutness of their desert ancestors, exist and can be picked out to perform a feat, and under the cloak of this exceptional solvency, is hid the all but universal bankruptcy of the race. If England ever wishes to test the real condition of her horses, let her cease to be satisfied with the performance of rare exceptions, and turn out a cavalry regiment or two, and as a trial assign them a task which any saddle-horse should readily perform; say, weight the horses with fourteen stone, march them daily 50 miles for a week, tying them up one night out of the seven without food or water. Give them a day's rest, then a field-day closed by a retreat of 70 miles. Those that go through the ordeal unscathed may be looked on as having gone through the trial of average horses ; and those that do not as useless.

The horse of England is tall of stature, and has a grace and beauty captivating to the eye of many, but which cannot deceive those who have much general experience of horses, and indeed becomes distasteful to them. They are soft and soft-looking. Take, as an instance, the horse which forms the frontispiece to Stonehenge, of whom he says, page 98, " He is thoroughbred, and in my belief can carry twelve stone against anything in this country. . . The likeness is most faithful, &c. Indeed, if we could obtain plenty of such horses, there would be nothing further to desire ; but he is an exception and can only be considered as the standard or type to be aimed at in breeding the hunter to carry twelve stone or fourteen stone." How long, let me ask, would a horse of his appearance last in war ? How would he stand the bivouac ? How would he stand the wear and tear of light cavalry practice before an enemy where forage might be scarce or bad ? How would he suit a bushman ? How would cold, or thirst, or hunger agree with this type of a hunter ? A thousand times rather give me the little Arab, at page 19, than this thoroughbred, whose appearance gives every indication of softness personified from head to hoof.

As compared with the Eastern horse that of England is costly in his food, a great eater and yet delicate in his appetite, a consumer of hay-stacks, and emptier of corn-bins; he is soft and washy; deteriorating on the turf, on the road a failure,—a national folly. How often have I heard Englishmen on dismounting from a ride in other countries exclaim "Well, how that fiery little devil carried me to-day! how he pulled and came in as fresh as paint! Why a horse of his stamp in England would have been jaded at the pace we went in twenty miles; only fancy fourteen stone on thirteen hands! and such roads; never made a false step, and a bolt in him to the finish. I suppose they're used to it; and then the food they get is nothing; and he'll be on his hind legs when I go to mount him in the morning."

That the Eastern horse in England is still an exotic, a hot-house plant, and has become degenerate and not acclimatized, he owes in a great measure to the Turf. As a rule in England the higher the breeding the more useless the animal; amongst the Eastern the purer the blood, the more capable the horse.

If the English horse promises little to the eye used to Arab horses, his performance is infinitely less. The Eastern horse has been transplanted to many soils; on English ground alone as a saddle-horse is he quite a failure.

The personal opinion of the writer of these pages may bring but little conviction to the reader; but the fact that (as far as I have been able to ascertain) there is no writer who, knowing personally the Eastern horses, and comparing them with those of England, does not give to the former a decided preference, for the fundamental qualities required in the saddle-horse, viz.: soundness, stoutness, vigor and endurance of hardships of all and every sort, is a circumstance which cannot be overlooked.

So much for the thoroughbred and saddle-horse of England, as gathered from personal observation and from the pages of Stonehenge and other English writers.

To amuse ourselves here in competing with the English Turf, is it not worse than useless? The Australian wants a good, sound, pleasant, fleet, hard-working, abstemious, hardy, and handsome saddle-horse. Has the English system through the agency of English races led to any such result either at home or abroad? I answer confidently on the authority of every author I have been able to meet, that for the last hundred years the saddle-horse of

England has shown a manifest, steady and progressive decline in all useful qualities. But if it is a most mistaken measure on the part of Australians, and contrary to the first principle we accept as our rule in breeding, to persist in breeding our saddle horses from the unfit, unsound, and delicate thoroughbred of England, this is not unfortunately our greatest error on the subject, which undoubtedly lies in the cultivation and acclimatization amongst us of the root and origin of this decadence, the adoption not of races, but of such races as have mainly wrought this damage at home, are perpetuating it here, and are practically a barrier to all amelioration.

CAVALRY HORSES.

CAPTAIN NOLAN ON "CAVALRY AND ITS TACTICS."

"Our Cavalry horses are feeble; they measure high, but they do so from length of limb, which is weakness not power."—*Nolan.*

From what has been submitted to the notice of the reader in the preceding chapter, I trust that his faith in the advisability of breeding saddle-horses from English racing sires will be somewhat shaken. But if such breeding is an error, it is not without many kindred ones. In fact a sort of fatality seems to have attended the management of horses by Englishmen and Europeans generally.

In the last 70 years we have done wonders for some at least of our other domestic animals. The produce of our sheep for instance, both in meat and wool, has been much increased in quality and bettered in description, and so in other things. We have gone on perfecting old sciences and creating new ones; thinking, weighing, and reasoning, we have gone on road making, steaming, manufacturing; disseminating our people and language on all sides; we have bettered and multiplied our material products; all that requires peace and security to grow has flourished. And so again in the art of destruction; our armies and volunteers have so increased in numbers and efficiency that it might almost be said, that the labor saved by steam to the hands of man, has been concentrated on the means of defence or aggression; but whilst our navy and artillery have armed themselves with a deeper thunder and a more deadly bolt, whilst our infantry has become infinitely more efficient than it once was, it would be difficult to show that our cavalry and their horses have not absolutely retrograded in every useful qualification. To make, however, the magnitude of our systematic mismanagement of saddle-horses, more unmistakably evident, I must bring forward further evidence on the subject and expose abuses and follies of another class. In the management and breeding of the thoroughbred horse in England, when his real purpose is considered, viz., no improvement in, or keeping up of the quality of the saddle-horses of that country, which does not for a moment enter into the purely commercial, and in no wise patriotic calculations of the breeder, but the production of an animal that shall stand a chance to fill his owners pockets by success on the race course, there is undeniably displayed an immense amount of experience and sound knowledge of the real means to attain the result sought. This I have never denied; what I have endeavoured to point out, is the extravagant folly of a system which assigns to these racers the office of sires for saddle-horses, in the face of every experience and of every rule the result of experience on this subject. I must now, however, go a step further, and show my reader that when the absolute necessity for proficiency in his undertaking, *that is the money test*, which is forced upon the racing man is removed, the skill which distinguishes him in his particular branch of horse management at all events, disappears likewise, and ignorance and folly usurp the places of knowledge and common sense. A remarkable exemplification of the truth of what I advance was found in the doings of George IV; nobody ever denied (for there is no denying racing results which are expressed but by two words, on which indeed the fate of too many depends—*success* or *failure*) his perfect capability for directing the management of his racers and racing stud, any more than his contemptible failure when he turned his royal attention to the management of cavalry.

Of no class of horses have we the same opportunity for forming a correct estimate as in that of the cavalry. Here we have many horses, treated on certain well authenticated principles, arriving at well known results. In turning to accounts of cavalry, one would expect to find, that however much the breeding of horses may have failed, that the management of such as have been selected for army purposes would have approached to something like perfection. Where the great national interests and honor of a people has so often fallen and will again in all probability fall to the lot of a few squadrons of cavalry to decide; where crowns, kingdoms, and colonies, where the fate of those who sit at the fire-side has been so often delegated to the arbitrament of the sabre; where so many intellects and for so long a time have been concentrated, we may suppose on that very important subject, the horse-soldier and his horse, it will be difficult to believe that anything but a judicious treatment, almost the perfection of horse management, has not long since become traditional in the service. With so much skilled labour at command, with the experience of ages to guide it, and the paramount importance of the subject itself intimately connected with the very being of a nation, here if anywhere we must expect to meet the results of a full and enlightened experience. As the remarks of a civilian on cavalry horses and their management might be looked down on as wanting the seal of professional knowledge to give them value, I will retire from the scene and allow the soldiers to speak for themselves. General F. de Brack, in his preface to his work on Light Cavalry, says, " War, said General Lassalle to me one day, is to the soldier who has not previously quitted his garrison, what the world is to the young man, who is leaving the forms of his school, *it is the application of the theory.*"

How does the British soldier and his horse stand this test? Lord Wellington, in his despatch 8th November 1810, says " Neither the Dragoons nor their horses are capable of performing much service, in the first year after their arrival, and many horses are lost, being unaccustomed to the food of the Peninsula and from want *of experience in the mode of taking care of them.*" So the theory fails in practice! and a new experience is needed. In "Cavalry, its History and Tactics, by Captain L. E. Nolan, 15th Hussars," we get a pretty clear insight both into the doings of our cavalry authorities as well as into the worth of our cavalry horses. In introducing this work to such of my readers as have not read it, it may be well to premise, that Captain Nolan had served in the Hungarian Cavalry, and was well acquainted with that arm in France, Austria, Russia, England, and India; was an enthusiast in his profession, by the members of which his book was well received. "The most hopeless condition to which an arm or science," says Nolan, "or an art can attain is that where its professors sit down with perfect self-satisfaction, under the conviction that it has reached perfection and is susceptible of no further improvement." A very proper and pertinent remark from an author who is just going to demonstrate, that the whole state of the arm of which he treats is rotten, root, stem, and branch! Speaking, for instance, of the Indian troops who did such good service, when fighting in their own fashion with their feet in short stirrups, and sharp blades in their hands, he says: "There is scarcely a more pitiable spectacle in the world than a native trooper mounted on an English saddle, tightened by his dress to the stiffness of a mummy, half suffocated with a leather collar, and a regulation sword in his hand, which must always be blunted by the steel scabbard in which it is encased."

I could multiply such instances and reflections, but will pass on to what chiefly concerns us, the horse. Lieutenant General Sir Charles Napier, as quoted by Nolan, says, "The hardships of war are by our dressers of cavalry thought too little for the animal's (the horse's) strength; they add a bag with the Frenchified name of *valise*, containing an epitome of a Jew's old clothes shop. Notably so if the regiment be hussars, a name given to Hungarian light horsemen, remarkable for activity, and carrying no other baggage than a small axe and a tea kettle to every dozen men. Our hussars' old clothes bag contains jackets, breeches of all dimensions, drawers, snuff-boxes, stockings, pink boots, yellow boots, eau-de-cologne, windsor soap, brandy, satin waist-coats, cigars, kid gloves, tooth-brushes, hair-brushes, dancing spurs; and thus, a *light* cavalry horse carries twenty-one stone. Hussars our men are not; a real hussar, including his twelfth part of a kettle, does not weigh twelve stone before he begins plundering."

"Without a system," says Nolan, "and a good system, it is impossible to make good troopers; at present we have none."

Many follies and mishaps, and amusing ones, might be transcribed from the pages of Captain Nolan, accounts of men armed with swords which will not cut, saddles on which it is impossible to sit, placed on the loins instead of the back, and girthed round the belly instead of the brisket, might be brought forward. All these, and a hundred other such absurdities, which are not accidental but parts of a system instilled into our soldier, and which send him to battle rather a victim than a combatant, might be adduced as proof of the empiricism to which the horse and all concerning him has been consigned. This, however, is apart from my subject, and I will at once hasten to place before my reader the results of Captain Nolan's experience of the stoutness and soundness of the English horse. The chapter is transcribed complete.

"Before I left India, some very interesting trials were made at Madras, by order of the Commander-in-chief, General Sir George Berkeley, the object of which was to test the capabilities of the troop horses, as well as the relative merits of entire horses and geldings for the purposes of war.

"Three trials were made. The first with two regiments of Native Regular Cavalry, one of stallions, one of geldings. The next with two troops of Horse Artillery. The third, and last, with two hundred English Dragoons (15th Hussars), one hundred riding stallions, and one hundred mounted on geldings.

"This squadron marched upwards of eight hundred miles, namely from Bangalore to Hyderabad, where they remained a short time to take part in the field days, pageants, &c. They then returned to Bangalore, four hundred miles, by forced marches; only one rest-day was allowed them, and the last six marches in were made at the rate of thirty miles a day. They brought in but one led horse; stallions and geldings did their work equally well, and were in equally good condition on their return. The question was, however, decided in favour of the latter, because they had been cut without reference to age, and only six months before the trial took place. The English cavalry in India is well mounted. On an emergency, any one of these Indian regiments would gallop fifty miles in a pursuit, leave few horses behind, and suffer but little from the effects of such exertion. The horses on which they are mounted are small but powerful. The Arab, the Persian, the Turcoman, the horses from the banks of the Araxes, are all unrivalled as war-horses. I have seen a Persian horse, fourteen hands three inches, carrying a man of our regiment of gigantic proportions, and weighing in marching order twenty-two and a half stone; I have seen this horse on the march above alluded to, of eight hundred miles, carrying this enormous weight with ease, and keeping his condition well. At the crossing of the Kistna, a broad, rapid, and dangerous river, the owner of this horse (Private Herne, of C troop,) refused to lead the animal into the ferry-boat to cross, but, saying, 'An hussar and his horse should never part company,' he took to the water in complete marching order, and the gallant little horse nobly stemmed the tide, and landed his rider safely on the opposite bank.

"An officer in India made a bet that he would himself ride his charger (an Arab little more than fourteen hands high) four hundred measured miles in five consecutive days, and he won the match; the horse performed his task with ease, and did not even throw out a wind-gall. The owner, an officer of the Madras Artillery, died shortly afterwards.

"General Daumas relates that the horses of the Sahara will travel, during five or six days, from seventy-five to ninety miles a day, and that in twenty-four hours they will go over from one hundred and fifty to one hundred and eighty miles, and this over a stony desert. Diseases of the feet and broken wind are almost unknown amongst them.

"What would become of an English cavalry regiment if suddenly required to make a few forced marches, or to keep up a pursuit for a few hundred miles! Their want of power to carry the weight, and want of breeding, makes them tire after trotting a few miles on the line of march.

"Our cavalry horses are feeble; they measure high, but they do so from length of limb, which is weakness, not power.

The blood they require is not that of our *weedy race-horse* (an animal *more akin to the greyhound, and bred for speed alone*), but it is the BLOOD OF THE ARAB and Persian, to give them that compact form and wiry limb, in which they are wanting.

"The fine Irish troop horses, formerly so sought for, are not now to be procured in the market. Instead of the long, low, deep-chested, short-backed, strong-loined horse of former days, you find nothing now but long-legged, straight-shouldered animals, *prone to disease* from the time they are foaled, and *whose legs grease after a common field day*. These animals form the staple of our remount horses.

"Decked out in showy trappings, their riders decorated with feathers and plumes, they look well to the superficial observer; *but the English cavalry are not what they should be*. If brought fresh into the field of battle, the speed of the horses, and the pluck of the men, would doubtless achieve great things for the moment; *but they could not endure, they could not follow up, they could not come again*.

"All other reforms in our cavalry will be useless, unless this important point be looked to. It is building a house on the sand to organize cavalry without good horses. Government alone could work the necessary reform by *importing stallions and mares of Eastern blood*, for the purpose of breeding troop horses and chargers for the cavalry of England.

"It is said that a Government stud is opposed to the principle of competition. What competition can there be amongst breeders for the price of a troop horse, when by breeding cart horses they obtain £40 for them when two years old? How could they possibly afford to rear animals with the necessary qualifications for a cavalry horse of the first class? To breed such horses a cross must first be obtained with our race-horses. This would entail a large outlay of capital, and when the good troop horse was produced, the breeder could not obtain his price for him.

"The rules of our turf encourage speed only, and that for short distances. *Horses are bred to meet these requirements, and from these weeds do our horses of the present day inherit their long legs, straight shoulders, weak constitutions, and want of all those qualities for which the English horse of former days was so justly renowned.*

"I had heard of fine horses in Russia, but I complacently said to myself, 'Whatever they are they cannot be as good as the English.' However, I went to Russia—and seeing is believing. Their horse-artillery and cavalry are far better mounted than ours, and their horses are immeasurably superior in those qualities which constitute the true war-horse, namely :—courage, constitutional vigour, strength of limb, and great power of endurance under fatigue and privation.

"The excellent example set by Sir George Berkeley, in India, might be followed up at home with great advantage to the service ; the capabilities of our cavalry horses of the present day should be severely tested, and the saddles should be tried, and experiments made to ascertain how sore backs may be avoided."

And yet how are we to reconcile this with the following assertion in an earlier part of his work, where he writes :—" I have heard it said that English horses are not adapted, like the Arab and other horses of Eastern breed, to skirmishing, to pulling up from speed, and turning quickly. The better the horse the more adapted to all feats of agility and strength. No horse can compare with the English, no horse is more easily broken into anything and everything, and there is no quality in which the English horse does not excel, no performance in which he cannot beat all competition."

This I at least put down as a rhodomontade which he has not failed to correct in his more serious mood when the subject came formally before him, and on several occasions. Nothing can be stronger or more contradictory than the conclusion of the former passage from Nolan. Here we have his estimate of the thoroughbred, of the English saddle-horse, and of the Arab. Stonehenge, too, invokes Government aid, and asks for sound and stout sires for the turf; Nolan calls for Eastern sires for the cavalry horses of England, and yet neither of them appear to me to have arrived at the pith of the subject, or to have seen the yet very obvious fact, that if "like begets like," that if the foal follows the sire, it is monstrous to expect good saddle-horses from racing sires, be those sires good or bad of their sort, and, of course, still less when we know how weedy, delicate, and unsound they really are.

For the amusement of the reader, I will add a few more particulars of cavalry and its management, from the same writer.

For example, he says, "When I was in India an engagement between a party of the Nizan's irregular horse, and a numerous body of insurgents took place, in which the horsemen, though far inferior in numbers, defeated the Rohillas with great slaughter.

"My attention was drawn particularly to the fight by the doctor's report of the killed and wounded, most of whom had suffered by the sword, and in the column of remarks such entries as the following were numerous, 'Arm cut off from the shoulder,' 'Head severed,' 'Both hands cut off (apparently at one blow) above the wrist, in holding up the arms to protect the head,' 'Leg cut off above the knee,' &c., &c.

"I was astonished. Were these men giants to lop off limbs thus wholesale? or was this result to be attributed (as I was told) to the sharp edge of the native blade, and the peculiar way of drawing it? I became anxious to see these horsemen of the Nizam, to examine their wonderful blades, and learn the knack of lopping off men's limbs. Opportunity soon offered, for the Commander-in-chief went to

Hyderabad on a tour of inspection, on which I accompanied him. After passing the Kistna River, a squadron of these very horsemen joined the camp as part of the escort. And now fancy my astonishment! The sword-blades they had were chiefly old dragoon blades cast from our service. The men had mounted them after their own fashion. The hilt and handle, both of metal, small in the grip, rather flat, not round like ours where the edge seldom falls true; they all had an edge like a razor from heel to point, were worn in wooden scabbards, a short single sling held them to the waist-belt, from which a strap passed through the hilt to a button in front, to keep the sword steady and prevent it flying out of the scabbard."

Again, says the same writer, " At ———, on the continent, Z Z shewed us the royal stables, and the horses broken in at the riding school. One of them had no shoes on; we asked the reason. Answer, 'He never works out of the riding school.' Question, 'How old is he?' A., 'Fourteen years old.' Q., 'Is he quite perfect in the riding-school work?' A., 'Not quite, but very good at it.'

"We were shewn a 'Springer.' A groom led in a horse with his tail tied on one side (I presume to give a better opening for the whip of the riding master), a cavesson on, and a young man in jack boots riding him, his legs drawn down and unnaturally far back, a cutting whip held upright in one hand, and the reins divided in both hands. The horse was placed against the side wall, the groom in front with the cavesson line held up to prevent the horse springing forward. The animal was evidently uneasy and looked back. No wonder! for presently the riding-master stepped up behind, and crack! crack! went the whip into the 'springer's' unprotected hind-quarters. He sprang in the air and back to his place, for he could not get forward. This was not enough. It appears that the perfection of this performance consists in getting the horse to kick out behind at the moment he is off the ground with all fours; and what between the groom pulling the iron band against the horse's nose with all his might, and the riding-master giving him the whip with a practised hand, he succeeded in getting the *capriole* required, sending the man in boots on to the horse's neck, at the same time. The riding master, pleased at the success of his experiment, turned to us to explain how difficult it was to get a horse to do it. I asked how long the horse had been at it. 'Oh,' said he, 'he has been a springer for several years. In fact, he was a lucky beast and had got his promotion early in life.'"

With these extracts from Nolan I will leave the reader to form his own idea about the management of cavalry horses, &c., by Europeans, and yet, as the same author remarks, "It doubtless requires great liberality and freedom from prejudice and preconceived opinion to admit that a system, on which the talent and experience of practical men has been exhausted for ages can be a bad one." May not these words be applied to many things in connection with our horses.

THE ARAB HORSE,

AND

THE HORSES OF THE SAHARA.

BY GENERAL E. DAUMAS.

" Rien n'est loin pour les chevaux."—*Proverbe Arabe.*

Having placed before the reader what appears useful for my purpose concerning the horses and horse-management of the English, I will now hasten to contrast him with a race of horses which, from many ages back, have been bred for the purpose of the saddle alone. A race of *pure saddle-horses*, in the real sense of the word. The progenitor of all the best saddle-horses which exist in the world—the Arab.

Amongst the various works treating of horses with which I have met, I know of none so likely to interest, amuse, and give information to the general reader as General Daumas's book, published about 10 years ago, entitled, " *Les Chevaux du Sahara et les*

Mœurs du Desert." In recommending it to the Melbourne reader—for a handsome edition of it presented by His Majesty the Emperor of the French is in the Public Library—I must warn him not to look for a repetition of many of our old English wonders, rules, or prejudices. Its faults and its merits are its own! It is fruit from a tree which does not grow in our gardens! To an Englishman interested in the horse, it is just such a treat as we might expect a German or Frenchman would experience in falling in with one of Nimrod's works for the first time. On what the author sees and hears, however, he makes but few remarks, and theorizes still less, remarking, "*Je ne viens nullement dire, ceci est bon, ceci est mauvais; je dis tout simplement, bon ou mauvais, voici ce que font les Arabes.*"

Besides a collection of facts, which by the generality of Australians will be found new, and the full record of a system in every way different from that of Europe, the author has succeeded, with the help of Abd-el-Kader, who largely contributed to his work, in throwing around the realities of his subject that garb of poetry and imagination with which the son of the desert loves to clothe even the sober realities of his every day life.

From the pages of Daumas I will now proceed to make translations of a few of such passages as appear likely to suggest matters for reflection, on the subject of which I write :—

"Horses," says Abd-el-Kader, "though all of one family, are of two sorts; the first is the Arab race, the other of the race of Beradin."

"A good horse," says Daumas, "ought to do 25 to 30 leagues (75 to 90 miles) five or six days consecutively; after two days of rest and good feed, he can begin the same work again. Journeys, however, in the desert are not always so long; but on the other hand it is common to see horses do 50 or 60 leagues (150 to 180 miles) in the twenty-four hours.

"On the subject of long distances traversed by desert horses, facts are quoted which would appear fabulous, were not the performers of them still alive, as well as witnesses who can confirm the truth of their statements. Here is one from a thousand, which was related to me by a man of the tribe of Arbâa. I will let him speak for himself:

"'I had come to the Tell, with my father and tribe, to purchase grain. It was under Aly Pacha. The Arbâa had had terrible quarrels with the Turks, and as it had become their interest, for the moment, to obtain by a show of complete submission, forgetfulness of the past, it was agreed amongst them, to gain over the immediate friends of the Pacha by money bribes, and to send to the Pacha himself, not as was their custom, a horse of medium qualities, but a steed of the first distinction. It was a misfortune, but it was God's will, and there was nothing for it but resignation. The choice fell on a mare of the colour of the grey-stones of the river, well known in all the Sahara. She belonged to my father who was warned to hold himself in readiness to start the next day and to take her to Algiers.

"'After the evening prayer, my father who had taken care to let fall no remarks on the matter, came to me and said, 'Ben Zyan, can I depend on thee to-day? wilt thou leave thy father in a dilemma? or wilt thou make his face blush?'

"'In me, sir,' I replied, 'there is only your will; speak, and if your orders are not obeyed by me, it will only be because I have been overtaken by death.'

"'Listen, then,' said my father, 'these children of sin, to settle their differences, wish me to take my mare to the sultan; thou knowest my grey mare, she who has always brought happiness to my tent, my children, and my camels; my grey mare, born the same day as thy youngest brother; speak, wilt thou allow them to defile my white beard with shame? The joy and happiness of my family are in thy hands; Mordjana (the name of the mare) has eaten barley (equivalent to being in high condition); if thou art my son, sup, take thy arms, and then at nightfall fly into the desert with the jewel that we all love.

"'Without saying one word, I kissed the hand of my father; I took the evening meal, and left Berouaguia happy in being able to prove my filial love, and laughing beforehand at the disappointment awaiting our Cheiks on awaking. I kept on for a long time, fearing that I was followed; Mordjana in the meantime pulling, and I trying to calm rather than to excite her.

"'Towards two o'clock at night I began to get sleepy, and pulling up alighted, and taking the reins rolled them round my wrist. Placing my firelock under my head, I lay down softly enough on one of those dwarf palm-trees so well known in our country. In an hour I awoke. Mordjana had eaten all the leaves of the dwarf palm-tree. We started. Daylight broke on us at Souagui; my mare had sweated and dried three times and began to take the spur. She drank at Sidi-bou-Zid in Ouad Ettouyl (Ettouyl Creek), and that evening I prayed the evening prayer at Legrouat, having given her a handful of straw to make her wait patiently the enormous nosebag of barley which she was to get.'

"These are not," said Si-ben-Zyan to me, "the journeys of your horses, for the horses of you Christians go from Algiers to Blidah, thirteen leagues!—*as far as from my nose to my ear!*— all the while thinking you are doing a long day's journey."

"This man," continues Daumas, "had done eighty leagues in twenty-four hours, his mare had eaten nothing but the leaves of the dwarf palm-tree on which he slept, she only drank once when half-way, and he swore to me by the head of the prophet that he could have gone and slept the next day at Gardaya (46 leagues further), if his life had been of danger. Si-ben-Zyan belongs to a family of Marabouts of the Oulad-Salahh, a branch of the tribe of Arbâa. He often comes to Algiers, will relate this adventure to any one who wishes to hear it, and will if necessary substantiate what he advances by credible witnesses.

"Another Arab of the name of Mohamed-ben-Mokhtar had come after the harvest to purchase grain at Tell; his tents were already erected on the Segrouan, and he was busy with his commerce with the Arabs of Tell, when the Bey Mezrag fell suddenly upon him at the head of a numerous body of cavalry, to punish one of those imaginary crimes, which the Turks so well know how to invent as a pretext for their rapacity. The attack had been made without any warning, and the foray was complete, the troopers of Makhzen giving themselves up to all the atrocities usual on such occasions.

"While this was going on, Mohamed-ben-Mokhtar hurriedly mounts his dark bay mare, a magnificent beast, known and envied by all the inhabitants of the Sahara, and impressed with the dangers of the occasion, he decides at once on sacrificing his whole fortune to the safety of his three children; one of them, four years old, he places on the pummel before him; another of six years of age behind him holding on by his belt, and the third he was about to deposit in the hood of his burnous, when he was prevented by his wife, who exclaimed, 'No, no, I will not give it to thee! They dare not kill an infant at the breast! Fly! I will keep it myself. God will protect us.' Mohamed-ben-Mokhtar then springs forward, discharges his gun and dashes from out the mêlée. Hard pressed he pushes on all day and the night following, and arrives the next evening at Laghrouat, where he is safe.

"A short time after his escape, he heard that his wife and child had been saved by some friends he had in Tell.

"Mohamed-ben-Mokhtar and his wife are still alive, and the two children whom he saved are known amongst the best horsemen of the tribe.

"What scene can be imagined more worthy of the painter's brush, than this of a family saved by a horse, from amongst pillage and bloodshed.

"And why should I seek to produce proofs of these facts? All the old officers of the division of Oran can relate how, in 1837, a general, attaching much importance to the obtaining of news from

Tlemcen, gave his own horse to an Arab to go in search of it. At four in the morning this man left Chateau Neuf, returning the next day at the same hour, having in the interim done 70 (seventy) leagues, through a country very much more rugged than the desert."

So much for the distances which can be done by the horses of the Sahara! And it should be borne in mind that these two instances selected from amongst a few of the like sort, in the first of which the horse accomplishes 80 leagues in twenty-four hours, and in the second does not much short of the same distance in thirty-six hours, having a man and two children on his back, are not related to us as the mere uncertain reports which might have been palmed off on a passing traveller. It must be remembered that Daumas, himself an old resident in the country, writes in the presence of thousands of French officers, who passed many years in Algeria, were often outmanœuvred and foiled in pursuit of men mounted on these very horses, and who had many excellent opportunities forced upon them of forming a correct estimate both of the capabilities and of the correctness of their historian.

But my reader must not imagine that "*Les Chevaux du Desert,*" is by any means a collection of anecdotes of this nature, they being, in fact, a material of which he has been very sparing, not indulging himself perhaps with half a dozen in the whole of his volume. Indeed, most readers, especially his own countrymen, would have relished a few more. But if he is sparing of prodigies, he is not so of customs and manners, of sayings and proverbs, of breeding the horse, feeding, physicking, breaking, shoeing, training, or bringing him up, &c.; on these, on the precepts of the Koran which apply to him, and many other points, his information is full, allowing nothing to escape him, except, perhaps, in the matter of the price at which he may be bought, of which he says not a word. In fact, with this exception, so thoroughly has he done his work, that but few new crumbs remain to be picked up by any one who would come after him. Besides this, he wrote under circumstances more fortunate than are ever likely to fall to the chance of another, for as he completed each chapter of his book, he submitted it to the judgment of Abd-el-Kader, then a prisoner in France. A remarkable position! not unlike one sitting with Pompey in the vessel that bears him to Egypt, listening to him speak of Mithridates and Pharsalia, before writing the fall of the commonwealth. A subject for the poet or the painter! Only fancy the great, the unfortunate emir; the scimitar-surrounded prince of the desert; the man of his day for a commentator! who, I may say, receiving into his hand a discussion, adorns it, etches it into pictures, throws on it the shadow of the burnous, strews flowers upon the page, and returns authenticated into the hand that gave him the dry truth, a poem rich with the imagery, the feel, the odour of his Eastern fancy—in matter a truth, in manner an ode! As I read the page I feel again in the East; the tent, the sand, the date-tree on the horizon, the Arab maiden coming from the well, the camel, the horse, the Slougui are again before my eyes, the odour of the chibouque is in my nostril, the musical gutteral of the Bedouin is again in my ear.

Then to what man of his time or of any other was the horse so much a stern reality as to Abd-el-Kader? For eighteen years he contended with the armies of France, backed by her immense resources. And what had he to oppose to this power? He was a living commentary on those words of the first Napoleon, "legs win more battles than arms." His *goums* bestrode horses whose locomotive powers, whilst defying pursuit, long annulled even calculations founded on the then supposed powers of the animal—to contend with whom, generals had to be educated anew, troops which might be called a new organization had to be set on foot to encircle this will-o'-the-wisp, enthral this ubiquitous foe, who so long opposed an army which has no superior, by the transcendent qualities of his horses alone. Well can we realize, as we read, his patriotic eulogium or the desert; his soldierly complaints of fortune, his manly regrets!

In this retouching of the subject by Abd-el-Kader, there is, however, nothing of the style of the commentator, far less of the reviewer; each tells the same tale after his own manner. The neat, concise, methodical, unadorned narrative of the son of France; followed by the brilliant, enthusiastic, richly imaged poem of the desert chief, redolent of the ardent fancies of the dwellers of the tent.

But if I cannot hope to give my readers any but a very general idea of a work whose beauty, like that of a statue, must be sought rather in its homogenous entirety than when broken in each disconnected limb, yet I shall, perhaps, be able to add from it to the foregoing some facts which bear upon my subject; facts which have brought irresistibly to my mind the conviction that horses, bred, reared, and treated on a system very different to that in vogue in England and in Australia, do still, in spite of all our prejudices, eclipse and surpass our own in every imaginable good quality except the one (if indeed it can be esteemed as a good quality) of speed for short distances alone. To put his assertion successfully to the test, would, if I am correct, require no careful culling of select specimens, but might with perfect safety be entrusted to any body of Sahara horses that might be found collected together for hunters or war-horses. I should like to see a thousand of the stoutest horses that could be produced in England, with fourteen stone on their backs, matched against a like number of Sahara horses, carrying a like weight, not for a day's gallop only, but for a month's march. I should like them to meet a little heat or cold, hunger or thirst, or all of them, on the road. I expect at the end of a month there would be a large hiatus between the two bodies, and that the result would go far to disenchant some of us Englishmen of our partiality for our very indifferent nags.

Amongst the circumstances in which the Barb is placed, no practice is begot more foreign to our ideas than that of his feeding. Barley alone, and that in very small quantities, is his most esteemed and often his only food. Touching this grain the Arabs of the Sahara have a host of sayings and legends. "He has eaten barley," is equivalent to "he is in training." "If we had not seen the mare give birth to the horse, we should have said he was born of barley." On a friend wishing to feed the horse of the Prophet, he is said to have exclaimed, "You wish then to have all my recompenses, for the angel Gabriel has told me that every grain of barley given by me to my horse will be counted as a good work for me."

In the spring his food often consists of the milk

of the sheep or camel, both of which are highly esteemed for this purpose. Dates, also, in various shapes, are often given to him; hay and grass are but little used, but chiehh, alfa, and one or two other plants that grow in the desert, are in high repute. Except a few mouthfuls in the morning, the Barb may be said to eat only once a day; sundown is about the hour he receives his 10 or 12 handfuls of barley. "The feed of the morning is found on the road," says the Arabs, "that of the evening on the croup of the horse." Water is given twice a day; in the morning it is limited to half a bucketful, and in the evening, about two hours before he gets his barley, he is allowed to drink his fill. But even this measure is at times still further curtailed, in order to give him the habit of abstemiousness, and enable him to bear thirst when necessary. For this purpose he is made to endure, yearly, two fasts during the height of summer, each of 40 days, during which period he is only watered once every second day. What would our consumers of haystacks, and emptiers of cisterns, with the thermometer at 120° in the shade, think of this?

Green-stuff is said by the Arabs to enlarge the stomach and fatten, and hence avoided, as the Arab of the Sahara likes his steed to be attenuated, and drawn up like a greyhound. They say, "The greatest enemies of the horse are rest and fat."

"On seeing," says Daumas, "two horses, one of the Tell and the other from the desert, a person who has not made himself well master of the facts of the case, would certainly prefer the former, looking so handsome, well-furnished, shining, and fat; whilst, misguided individual, he would despise and calumniate the second for those very qualities in which consist all his worth; for his fine, hard, and dry legs, his drawn-up stomach, and his bare ribs; and yet this desert horse who has had but little barley, knows not green-stuff or straw, whose feed has been chiebh, bouse, and seuliane, whose drink has been milk, and has from his youth run the chase and the raid, has the swiftness of the gazelle and the patience of the dog, and by his side the horse of the Tell is but an ox."

Of breeding, Daumas and the Emir say much. The Saharians hold the sire to be of much greater consequence than the dam, which latter in theory, perhaps, though not in practice, they undervalue. In the choice of a sire they are most particular, going any distance to obtain one, and allowing 2 or 3 years to pass rather than accept one that is not approved of. Not more than six or eight mares are allowed to a horse in one season; his services are gratuitous, one who should take money for them, say they, "would be called *marchand d'amour du cheval.*" Neither is the mare kept constantly breeding, as such a course is said to weaken the offspring. Good and bad qualities of every sort, injuries from work, unsoundness, &c., are said to re-appear in the offspring, hence a sire of repute always excels in such qualities as are sought in the desert.

The "breaking" of the Arabs, though gentle, is prolonged and severe, as they exact more acquirements from their horses than we do from ours. It begins very young—at 30 months old. "Mount him," say they, "between two and three years of age, and ride him gently; tie him up from three to four, and feed him well; then remount him, and if he does not suit—sell him." He is said usually to live five and twenty years, and with respect to the period of his utility, the proverb is:—

Seven years for my brother,
Seven years for my self, and
Seven years for my enemy.

Whatever may be the result of such early breaking, there is no doubt but that it finds great favour with the Arabs. Besides common testimony, and that of Abd-el-Kader on the subject, Daumas brings forward that of the Kalif of Mekjana, chief of one of the most illustrious families of Algeria, who says: "During my long career, I have seen reared by my tribe, friends, and servants, more than 2,000 colts, and I affirm that all those whose breaking was not begun early, and on the principles contained herein, have always turned out intractable, disagreeable, and useless for war. I further affirm that when I have made long and rapid marches at the head of 1,200 or 1,500 horsemen, that the horses without fat, thin even, but early used to fatigue, have always stuck to my standard, whilst those which were fat, but had not been broken whilst young, were always left behind.

"My conviction on this head is the result of such full experience, that, when being lately at Cairo, and requiring to purchase some horses, I refused without pity all such as were offered to me that had not been early broken.

"'How has thy horse been brought up?' was always my first question.

"'Sir,' replied an inhabitant of a town, 'this horse has always been reared like one of my own children; well fed, well looked after, and not overworked, for I did not mount him till he was four off. See how fat he is, and how clean his legs!'

"'Very well, my friend, keep him; he is thy pride, and that of thy family; it would be a shame to my grey hairs to deprive thee of him.'

"'And thou,' said I, turning to one whom I knew, so dark and sunburnt was he, to be a child of the desert, 'how has thy horse been reared?'

"'Sir,' said he, 'early I accustomed his back to the saddle and his mouth to the bit; with him I have often fallen on my foe who dwelt far, very far off; many are the days he has passed without water, and nights without food; his ribs in truth are bare, but if you are waylaid on the road he will not leave you in trouble. This I swear by the day of judgement, when God will be Cadi, and his angels witnesses.'

"'Picket the grey before my tent,' I would say to my servants, 'and satisfy this man.'"

Daumas gives us also a pretty full account of the turf meetings in the desert, on which gambling and betting are forbidden, the horses only running for honor and the stakes. Forty days is the period used for training and reducing the horse, after he has been first allowed to fatten gradually on barley. The prophet, who appears to have regulated all the affairs of his people, has also fixed the limits of their races. To cock-tails he allots a course of one mile, and to pure horses seven miles; he does not appear to have made any provision for a Jockey Club, (indeed how could it exist without betting?) or, probably all this would have been altered long since; the distances, no doubt, being curtailed very much, and the horses not ridden as they now are by men, but to lighten their loads, by monkeys.

Colour in horses is regarded in a much more

serious manner by the Arabs than it is by us, and according to my experience, with good cause. All light and ill-defined colours are despised as soft, whatever the build of the animal may be. Very dark chesnut is thought to be the colour usually borne by fast horses, and together with dark bay and black, is most esteemed for speed, courage, and bottom. Grey, which is supposed by us to be the Arab color, is esteemed handsome, but not thought to denote good stuff; " in the sun," say they, " he melts like butter, and in the rain like salt," which I think is quite true. The dun and roan are also held to be very inferior, and of the piebald, they say, "avoid him like the plague ; he is brother to the cow."

On the subject of the "evil eye," and the forty signs of good and bad augury, the general gives us ample imformation. The latter superstition has, at least, one happy result, which is, that horses of rare excellence may sometimes be bought cheap, as I can bear testimony, whose only fault is one of those unpopular marks, which would prevent any Arab of distinction from riding him, as he is esteemed unlucky, and sure to bring his rider to grief, sooner or later.

The great test of perfect formation in a horse and that his proportions are as they should be, is his being able to drink from the level of the ground without bending either of his knees, keeping his legs firm beneath him. The mountain horse is preferred to that of the plain, and the horse bred on swampy ground is thought only fit for the packsaddle, "*prefère le cheval de montagne au cheval de plaine, et celui-ci au cheval de marais, qui n'est bon qu'à porter le bât.*"

Daumas in his book bears testimony to the rare qualities of the Sahara horses, and thus speaks of them in his *Avant Propos* to the Fourth Edition. " In the Crimea, where we had four magnificent regiments of *Chasseurs d'Afrique* mounted on the Barb horse, he was well tried, and if, as the song says, he could ' endure hunger and thirst,' he can also, which has been strongly denied, endure exile and the rigours of a cold and wet climate." Again, comparing them with the horses of France, he says, "The European horse has disappeared from our army in Africa, of which he could bear neither the impetuous charges nor the incessant marches. He has been replaced by the horse of the country. The first care of an officer coming from France is to furnish himself with horses bred in the country. He takes good care not to venture into the desert, still less into the mountains, with such horses as would be very much esteemed at Chantilly, on the Champ-de-Mars, or at Sartory." (Where, in fact, English horses are most admired and much ridden.)

Having thus touched concisely on what is principally interesting, and what bears on my subject in "*Les Chevaux du Desert,*" I will leave my reader to form his own conclusions as to the value of the work, and the soundness of the principles and practices advanced in it, only adding that so highly do the Arabs esteem their own horses, that the admixture of any foreign blood is accounted a degradation, and that, when the Emir was in power, the sale of a horse by an Arab to a Christian was punished with death.

Since then, however, he appears to have somewhat relented in this particular, as it is not long since he presented to his majesty, Napoleon III., three celebrated horses, carefully chosen from amongst thousands, and which were accompanied by a letter thus translated :—

"Ever since my residence was fixed at Damascus I have never ceased to look out for stallions for the emperor's stud, which, in point of purity of blood and those other qualities which my experience and special knowledge enable me to judge of, should come up to the standard of my ideas. Whenever I heard of a stallion among the tribes bounded by Egypt, Irak, and Hedjex, I never failed to make the most particular inquiries about him. I was long, however, before finding any horse combining that purity of blood and perfection of form and qualities which I had set my mind upon. I went myself among the Arab tribes of Nedj and Mesopotamia, at the time when they were in the neighbourhood of Damascus, but could not meet with what I wanted. At length, however, I have found a dark bay stallion with black mane and tail, rising seven years old, such as is not to be matched in all the tribes of the desert. His reputation is so great that the Bedouin Arabs would take twenty days' journey, and more, to take a mare to him. I have besides selected a three-year-old colt, who, for beauty and noble race, is unequalled. The Arabs can trace his family for fifty generations, pure from any crossing. Desiring to present these horses to his majesty, I beg your excellency to send the necessary orders to the French consul at Beyrout to have them embarked with the grooms in charge of them, in order that they may be taken to his imperial majesty. Salutation on the part of your devoted Abd-el-Kader, son of Mahhy-Eddin. ABD-EL-KADER."

The following description of these horses I have cut out of the Melbourne *Argus* :—

" ABD-EL-KADER'S PRESENT TO THE EMPEROR OF
THE FRENCH.

" The following letter appears in the *Field* :—

" Sir,—A charming drive through the Bois de Boulogne takes one to the Port St. James, outside which is the new Imperial stud establishment, to which the public are admitted from one to five o'clock daily, without cards. The visitor will see there the Flying Dutchman, Fitz-Gladiator, and other horses of less notoriety, as also the three Arabian horses that were given by Abd-el-Kader to the Emperor.

"Those Arabians, as you may suppose, are the attraction of the day; and no wonder that they should be so, when we learn, from the extracts which have been published of the ex-Emir's letter, all the time and trouble it has cost him to select those three high-bred horses. To get possession of one of them (Emir) he spent four years in the desert amongst the tribes; and when we consider the value those people set upon their best horses, and examine this horse attentively, we become impressed with the notion that he must have cost a fabulous price—perhaps his weight in crowns; or, if paid for in lands, they might represent a territory as large as a kingdom; but, whatever the price, the gift is worthy of Abd-el-Kader, and the thanks of the Emperor. I do not suppose the equal of this horse, as a high-caste Arabian, has been brought to this country or England from the desert in our day. How his stock may turn out, time alone will tell; but I shall be very much disappointed if he does not materially improve the breed of horses in this country, if he only gets good mares. In a paragraph, as translated to me, of the ex-Emir's letter, he speaks in the language of the desert; it is so remarkable that I give it to you word for word. He says, 'Good horses are like friends; a man meets with only a few in his lifetime. As one man is worth a thousand other men, so one horse is worth a thousand others; and a thousand bad horses are not worth a single good one.' This horse Emir, they say, proved himself a good one at all distances. He is of the Kohel-Obaian race, was bred in Trac; his pedigree is known to be of the best and purest blood for twenty generations. He is eight years old, colour brown, with a very silky skin, three white fetlocks, some white, but not too much, about his face. He has some white spots about the size of a small pea on his neck and shoulder. He stands fourteen hands one inch high, has a good, long, lean head, well set on, ears slender, his eye mild and intelligent, his mane and tail light and the hair fine, a handsome straight strong neck, longer than one generally sees with other Arabs; his shoulders are long and well laid in, each of those points showing much quality. He has great legs and quarters, ribs and back wide to a degree; hoofs black, strong, wide at the heels, and deep; joints large; knees and hocks very good. He is all over a remarkably strong, thick-made horse, but to my taste, too short to be speedy, and sinks a little too much upon his fetlocks, but we cannot have perfection; his action is more elastic than might be expected from a horse of his build and strength. He is as gentle as a sheep without being dull; he looks like what we would call a craving but very aristocratic animal.

"Ismael is five years old, a very dark brown, two hind fetlocks white, stands fourteen hands one inch high; has a fair shaped head, forelegs rather light, feet good and wide, quarters long, back level, loins free, shoulders plain and somewhat thick, mane and tail heavy and coarse. Taking him altogether, he is a loose, lengthy-looking horse, but not showing much quality.

"Moor is four years old, chesnut, two fetlocks (near side) white, stands fourteen hands high; he will grow another inch. His head and neck are stiffly set on; his shoulders, legs, and feet are good, so are his back and ribs. His hocks are also good, but he sinks too much upon his fetlocks. His action is gay and free; he is a wiry-looking colt. Abd-el-Kader informs us that there is no horse in the desert whose blood is purer than that of the Moor, and no pedigree better known for fifty generations. His condition is very low, he appears to have suffered much more than his companions upon the passage. He is much cut up, but when he recovers and is a year older, I make no doubt he will have many admirers.

"All the horses are very gentle; no biting or kicking out when they are handled by the grooms or visitors, which is too often the case with such horses under our management. Why it should be so I am at a loss to conjecture; could it be that we bully them with our voice and action! I observe the French grooms are always teaching their horses some simple equestrian tricks; this leads to a bit of sugar or a carrot as a reward.

"Paris, June 10. "FARMER."

But the part of *"Les Chevaux du Desert"* which is most easily separated from the rest, as well as that which will perhaps be most interesting to the reader, is the Emir's celebrated letter to General Daumas, which, with its introduction, I have translated as follows:—

"Having known the Emir Ald-el-Kader whilst I was French Consul at Mascara, from the year 1837 to 1839, and having again met him at Toulon, when sent there on service, at the period of his arrival in France, I have been able, in my numberless interviews with him, to appreciate his profound knowledge of everything relating to the equine history and science of his country. This has led me without hesitation to ask his opinion on matters which, though purely scientific, embrace nevertheless, features of very great interest, not only to the future of our Algerine colony, but to the very capital of France as well.

"Annexed is the letter which I received in reply from him, dated 8th November, 1851, (23rd of Moharnem, first month of 1268.)

"'Glory be to the one only God—His reign is alone eternal.

"'Greetings to him who equals in good qualities any man of his day, who seeks but good, whose heart is pure, and who keeps his word, to the wise intelligent and illustrious General Daumas, from his friend Sid-el-Hadj Abd-el-Kader, son of Mahhi Eddin.

"'Below are my answers to your questions.

"'1st. *You ask me how many days the Arab horse can travel without rest or becoming too much jaded?*

"'Know then that a perfectly sound horse, having as much barley as he needs, will do all that his rider may require of him. The Arab saying on this subject is—

Allef ou Annef,
Donne de l'orge et abuse.—DAUMAS.

"'Fill him with barley and ride him as you like. But without what may be termed abusing your horse you may ride him 16 parasangs daily (60 miles). This is the distance from Mascara to

Koudiat Aghelizan on the Mina river, and has been measured in drâa (cubits). A horse doing this distance every day, and having as much barley as he requires, can keep on at this task without being overworked, for three or even four months without resting a single day.

"'2nd. *You ask me what distance can a horse travel in a single day?*

"'This I cannot tell you precisely, but the distance should be about 50 parasangs (187 miles), as from Tlemen to Mascara. But the horse that has performed this journey should be spared the day following, on which he would be able to perform a much shorter distance. The generality of our horses used to go from Oran to Mascara, and would have been able to repeat the performance two or three days following. We once started from Saïda at 8 o'clock to fall upon the tribe of Arbâa which was encamped at Aaïn Toukria (in the country of the Oulad Aïad near Taza) and we reached it at daylight. You know the country and the distance we should have to perform.

"'3rd. *You ask me for instances of abstemiousness in the Arab horse and for proofs of his power of enduring hunger and thirst.*

"'Know then that when we were encamped at the mouth of the Melouia, we made hostile excursions into the Djebel Amour by way of the Sahara, push-

F

ing our horses on the day of the attack for five or six hours without drawing bridle, accomplishing our inroad, going and returning in 20 or at most 25 days. During this period our horses had no barley but such as they could carry with their riders, say about eight ordinary feeds; neither had they straw, but only alfa and chiehh joined to a little grass in spring. Nevertheless on our return to our people we were caracolling our horses on the very day of our arrival, and even performing festive caprioles with a certain number from amongst them. Many, however, which could not have gone through this latter exercise were yet fit for active service.

"'Our horses could remain one or two days without drinking, once even passing three days without obtaining water. But the horses of the Sahara do still more than this. They remain three months without tasting barley. They never taste straw except when we come to the Tell to buy grain, and in general eat nothing but alfa and chiehh, or sometimes quetof; chiehh is preferable to alfa, and quetof to chiehh. The Arabs say—

> *Alfa fits for the march,*
> *Chiehh for the fight,*
> *And quetof is better than barley.*

"'Some seasons when the tribes have not been received into the Tell, a whole year passes without the horses of the Sahara tasting one grain of barley. On such occasions dates are sometimes given the horses. This food fattens them and enables them both to fight and to march.

"'4th. *You ask me how it is that though the French only break in their horses as four-year-olds, the Arabs ride theirs whilst very young.*

"'Know then that the Arabs say, that the horse, like man only learns easily whilst young; their proverb on this head is—

The lessons of childhood are graven on stone; the lessons of maturity disappear like birds' nests.

And again—

The young branch can easily be straightened; the thick limb can never be made straight.

"'In his first year the Arabs teach the horse to lead with the *reseum*, which is a sort of cavesson. At that age they call him djeda and begin to tie him up and bridle him. When he comes to be *teni*, that is, when he enters his second year, they ride him a mile, then two, then a parasang, and when he is 18 months old they no longer fear to tire him.

"'When he becomes *rebâa telata*, that is, when he enters his third year, they tie him up, refrain from riding him, cover him up with a good cloth *(djelale)*, and fatten him. Of this practice they say—

> *In his first year (djeda) tie him up so that no accident befall him. In his second year (teni) ride him till his back bend. In his third year (rebâa telata) again tie him up, then if he fails to please you, sell him.*

"'If a horse is not ridden until his third year, it is certain he will be good for nothing, unless it be to gallop, for this is not to be taught, it is the original faculty, on which subject the Arabs say—

El djouad idjri be aâselouh.

The djouad gallops in accordance with his blood. (The horse of noble blood gallops without teaching.)

"'5th. You ask why our mares are more costly than our stallions, if it be true, as we affirm, that the foal inherits more of the qualities of the sire than of the dam?

"'The reason is this, he who buys a mare hopes, whilst he rides her, to have many foals from her; whilst he who purchases a stallion can only ride him; *for the Arabs make no charge for the use of their sires, but lend them free of cost for stud purposes.*

"'6th. You ask me if the Arabs of the Sahara keep

regular registers as guarantees of the pedigrees of their horses.

"'Know then that the people of the Sahara take no heed of such registers any more than do those of the Tell. Notoriety is a sufficient guarantee for them. The pedigrees of their pure horses is as well known to them as is their own. I have heard, it is true, that some families have written genealogies, but I cannot point them out. Such books, however, are in use in the East, as I mention in the little treatise which I am about to dedicate to you.

"'7th. *You ask which of the Algerian tribes are the most celebrated for their noble horses.*

"'Know then that the best horses of the Sahara are, without exception, those of the Hamyane. They have none but good horses, as they keep none for draft or packing, using them only for expeditions and the battle-field. It is these horses that best support hunger, thirst, and fatigue. After the horses of the Hamyane come next in quality those of the Harar, of the Arbâa and the Oulad Nayl.

"'In the Tell the horses most distinguished for noble blood, height and beauty, are those of the people of Chelif, principally those of Oulad-Sidi-Ben-Abdallah (sidi-el-Aaribi), near the Mina, as well as those of the Oulad-Sidi-Hassan, a branch of the tribe of Oulad-Sidi-Dahhou, who dwell in the mountains of Mascara. The fastest on the race-course, and of beautiful figure, are those of the tribe of Flitas, the Oulad Cherhif and the Oulad Lekrend. Those which travel best on rocky ground without shoes are the horses of the Assassena, in Yabroukia. To Moulaye-Ismaïl, the celebrated Sultan of Morocco, is attributed this saying—

> *May my horse have been fed in Mâz,*
> *And watered in the Biâz.*

The Mâz is a country belonging to the Assassena, and the Biâz is a brook, which under the name of Foult runs their territory.

"'The horses of the Oulad-Kaled also are renowned for these same qualities. Sidi-Ahmed-Benoussel has said on the subject of this tribe—

> *Long tresses and long djelals shall be found*
> *with you even to the day of resurrection.*

Thus passing a eulogium on their women and their horses.

"'8th. *You tell me that the horses of Algeria are said not to be Arab but Barb horses.*

"'This is an opinion that recoils on its authors. The Barbary tribes are of Arabic origin. A celebrated author has said—

> *The Barbs inhabit the Mogheb; they are all*
> *sons of Kaïs-ben-Ghelan. We are assured*

that they come of the two great tribes of the Hemearites; the Senakdja and the Kettama came into the country at the epoch of the invasion of Ifrikech the Malek.

" 'After these two opinions there is no gainsaying that the Barbs are Arabs. The historians, besides, give the genealogy of the Barbary tribes, and trace their descent from the Senahdja and Kettama tribes. The advent of these tribes was anterior to Islamism. Since the Mussulman invasion, the number of Arabs who have emigrated to the Mogheb is incalculable. When the Obeidin (the Fatimites) were masters of Egypt, immense tribes went on to Africa, amongst others the Reahh. They spread themselves from Kaironan to Morocco. From these tribes are descended in Algeria, the Douaouda, the Aiad, the Maded, the Oulad-Madi, the Oulad-Ihaonb-Zerara, the Djendel, the Attof, the Hamis, the Braze, the Sbeha, the Fleta, the Medjahar, the Mchal, the Beni-Amer, the Hamian, and many others. There can be no doubt but that the Arab horse was disseminated with the Arabs themselves. In the time of Ifrikech-ben-Kaif, the Arab empire was all powerful ; it extended on the west to the furthest outside limits of the Mogheb, as in the days of Chamar the Hemiarite, it reached as far as China, as Ben-Kouteiba relates in his book called *El Marif.*

"'It is true, however, that if all the horses of Algeria are of the Arab breed, that many, from being used for draft, harness, the pack-saddle, to drag burdens, and other such work, and from the mares having been put to the ass, none of which things were formerly done amongst the Arabs, that they have fallen from their pristine nobility. On this subject, it is said, that it is enough for a horse to walk even on ploughed ground to lose some of his worth. On this subject the following story is traditional:—

"'A man was riding a horse of noble breed. He met his enemy mounted as well on a noble steed. One of them pursued the other, and the pursuer was distanced by the pursued. Hopeless of overtaking him, he shouted out at the top of his voice,

> *I ask thee, in the name of God, has thy horse ever been in the plough?*
> *He has ploughed for four days.*
> *All right! Mine has never ploughed. By the head of the Prophet I am sure to overtake thee.*

The chase continued. Towards sundown the fugitive began to lose ground, and his pursuer to overtake him, and at last closed with him whom he had despaired of catching.

"'My Father, God have mercy on him, was accustomed to say there has been no blessing on our *country since we have packed and harnessed our horses. Did not God create the horse to ride, the ox to plough, and the camel to carry the pack? There is no gain in swerving from the paths of God!*

" '9th. *You ask me what are our precepts touching the feeding and keeping of our horses in condition.*

" ' Know then that owners of horses begin with giving their horses at first but little barley, increasing the feed gradually, then diminishing it again until the horse leaves a little of it uneaten. He is then restricted to this quantity.

" ' The best time to feed is the evening. Except when travelling, it is useless to feed in the morning. It is a saying on this subject—

The barley of the morning is found in the dung,
The barley of the evening re-appears in the croup.

" ' The best way to feed with barley is when the horse is saddled and girthed, as it is best to give him water with his bridle on.

" ' On this subject we say—

Water with the bridle,
Barley with the saddle.

" ' The Arabs place great store on a small-feeding horse, provided he be not weakened by it. *He is,* they say, *a priceless treasure.*

" 'Water in the morning makes a horse lose flesh,
" 'Water in the evening fattens,
" 'Water at mid-day preserves condition.

" 'During the great heats, which last 40 days, the Arabs only water their horses once every two days. They say that this custom produces the best results. In summer, autumn, and winter, they give an armful of straw to their horses; but their chief article of food, as well as that which they prefer to everything else, is barley.

" 'The Arabs on this subject say—

If we had not seen that horses spring from horses, we should have said, it is barley that begets them.

" 'They say,

Find a large horse and buy him, barley will make him go.

" 'They say,

Of forbidden meats choose the lightest.

That is, choose a light horse; horseflesh being a forbidden food to the Moslems.

" 'They say,

One becomes not a horseman till after many falls.

" 'They say,

The pure horse has no vice.

"'They say,

> *The horse in readiness is the honor of his master.*

"'They say,

> *Horses are birds without wings.*

"'They say,

> *Nothing is far for the horse.*

"'They say,

> *He who overlooks the beauty of the horse for that of women will never prosper.*

"'They say,

> *Horses know their riders.*

"'The holy Ben-el-Abbis, may he find favour with God, has said also,

> *Save your horses and have a care of them,*
> *Spare not your attentions,*
> *On them depend honor and beauty.*
> *Were horses abandoned by men,*
> *I would take them into my family,*
> *And share with them the bread of my children.*
> *My wives clothe them with their veils,*
> *And cover themselves with their djelales.*

Every day I lead them
To the field of adventure.
Borne forward in their impetuous course,
I do battle with the most valiant.

"'I have completed my letter, which our brother and companion, the friend of all, the Commandant Sid-bou-Senna, will cause to be put into your hand. Greeting.'

"This letter was written by the hand of Abd-el-Kader, the original is still in my possession, &c."—Daumas.

THE ARAB HORSE

IN INDIA, SYRIA, MESOPOTAMIA, AND PERSIA.

"Veritable buveur d'air
Il noircit le cœur de nos ennemis."—*Daumas.*

It is probable that the reader will turn but a deaf ear both to Abd-el-Kader and Daumas' estimate of the Arab horse. For this reason I hasten to place before him the views of some English writers, who have had good opportunities of becoming acquainted with and testing his qualities in foreign lands; and let me remark I have met with no rebutting evidence to the facts and opinions which are advanced by these writers. All Englishmen who have really tried the Arab at work, seem to come to one conclusion—that he has no equal. Let us begin with the Arab horse in India.

Of the Arab in India I know but little personally, never having been in that country. I have seen about twenty Indian Arabs in Australia, which were said to be, and as they were brought here as sires would probably be favourable specimens of their class. These few specimens had all a decided family resemblance, and appeared to me weedy when compared with the Arab of Syria. Their descendants here have not added to their fame in Australia. From these censures I must except "Satellite," whom I never saw, but whose stock, twenty-five years ago, were in high repute in New South Wales. Descriptions, however, of the Indian Arab, by writers who knew him well, are numerous, and his characteristics well determined. None that I have met with are drawn so graphically, and by such a master hand, as that which is to be found in the supplementary chapter to the "Wild Sports of India," by Captain Shakespeare, which is headed—"On the different breeds of horses in India."

"In the first place," says Captain Shakespeare, "the Arab horse is the very best horse under saddle, for all general purposes, that can be had in India. If anything besides general opinion is required to corroborate this, it is found in the fact, that in the market the Arab horse invariably commands the highest price, whether he is bought for a racer, a charger, a hunter, or a hack. *He is the soundest horse, the most enduring, the most beautiful to the eye, the most docile, and the most courageous, and*

THE ARAB HORSE. 111

he is more easily broken in than any other. . . .

. . . Bombay and Bangalore are the chief marts for the Arab horse, and in the stables at the former place, from November to February, you may see as many as a 1,000 fresh horses for sale. Out of these, 50 perhaps are high caste horses, either Nedjd—which is the pure Arab: rarely standing above fourteen, and more commonly only fourteen hands, or under—or the Aneezah Arab: the highest form of which is bred by a tribe of Aneezahs that inhabit the Desert, some two marches from Bagdad. . . .

. . . It is worthy of notice, that with reference to carrying weight, the Arab horse runs in exact contrast to the thorough-bred English, *whose best blood is derived from him.* . . . . . In India, the low Arab horse cannot compete with the taller horse at heavy weights, either on a race-course or across country: thus we have many races weight for inches. To look at these low sturdy built horses, you would think that they could gallop under any weight, but the trial will undeceive you. . . . . On the whole the chesnut and dark chesnut are the most courageous. The dark coloured are also not fiery, and they are also more generally weight carriers. There are more grey Arab racers than of any other colour."

The following picture of the Arab horse is so interesting, and drawn with so masterly a hand, that

I insert it without excuse, though not bearing directly on the subject:—

"The points of the high caste Arab, as compared with the English thoroughbred, are as follows: The head is more beautifully formed, and more intelligent; the forehead broader; the muzzle finer; the eye more prominent, more sleepy looking in repose, more brilliant when the animal is excited. The ear is more beautifully pricked, and of exquisite shape and sensitiveness. On the back of the trained hunter the rider scarcely requires to keep his eye on anything but the ears of his horse, which gives indications of everything which his ever-watchful eye catches sight of. The nostril is not always so open in a state of rest, and indeed looks often thick and closed; but in excitement, or when the lungs are in full play from the animal being at speed, it expands greatly, and the membrane shows scarlet, as if on fire. The game-cock throttle—that exquisite formation of the throat and jaws of the blood-horse—is not so commonly seen in the Arab as in the thoroughbred English race-horse; nor is the head quite so lean. The jaws, for the size of the head, are perhaps more apart, giving more room for the expansion of the wind-pipe; the point where the head is put on to the neck is quite as delicate as the English horse: this junction has much more to do with the mouth of the horse than most people are

THE ARAB HORSE.

aware of, and on it depends the pleasure or otherwise of the rider.

"The bones, from the eye down towards the lower part of the head, should not be too concave, or of a deer's form: for this in the Arab, as in the English horse, denotes a violent temper, though it is very beautiful to look at. Proceeding to the neck, we notice that the Arab stallion has rarely the crest that the English stallion has. He has a strong, light muscular neck: a little short perhaps compared to the other, and thick. In pure breeds, the neck runs into the shoulder very gradually; and generally, if the horse has a pretty good crest, comes down rather perpendicularly into the shoulder; but often, if the horse is a little ewe-necked, which is not uncommon with the Arab, it runs too straight and low down into the shoulders. The Arab horse, however, rarely carries his head, when he is being ridden, so high in proportion as the English. He is not so well topped, which I attribute to the different way he is reared, and to his not being broken in regularly, like the English horse, before he is put to work. His shoulders are not so flat and thin, and he is thicker through in these parts generally, for his size, than the English thoroughbred horse. His girth does not show so deep, that is, he does not look so deep over the heart; but between the knees and behind the saddle, where the English horse very often falls off, the Arab is barrel-ribbed, and this gives his wonderful endurance, and his great constitutional points. This also prevents him getting knocked up in severe training, or under short allowance of food, and in long marches.

"His chest is quite broad enough and deep enough for either strength or bottom. The scapular or shoulder blade, is both in length and backward inclination, compared to the humerus or upper bone of the arm, quite as fine in the high-caste Arab as in the English horse, while both bones are generally better furnished with muscles, better developed and feel firmer to the hand. But some of the very fastest Arabs have their forelegs very much under them: so much so, indeed, that no judge would buy an English horse so made; yet, whether it be that this form admits of the joints between these bones becoming more open when the horse extends himself, or whatever be the cause, it is a fact that blood horses thus made, are almost always fast horses. The upper part of their shoulder blade seems to run back under the front part of the saddle, when they are going their best.

"This formation is most common in the lower-sized Arab, and apparently makes up for his deficiency in height: the very finest Arabs have this peculiarity of form. They are rather apt to become chafed at the elbow-points by the girths, and almost require to have saddles made for them. The elbow-point, that essential bone, which, for the sake of leverage, should be prominent, is free in the Arab, and generally plays clear of the body. The forearm is strong and muscular, and is pretty long; the knee square, with a good speedy cut for the size of the animal, equal to the English horse; while below the knee, the Arab shines very conspicuously: having a degree of power there, both in suspensor ligaments, and flexor tendons, far superior in proportion to his size to the English horse. These are distinct and away from the shank bone; they give a very deep leg, and act mechanically to great advantage.

"The bone looks small, but then it is very dense: the hollow which contains the marrow being very small, and the material solid, more like ivory than bone, heavy and close-grained. The flexor tendons are nearly as large and thick as the cannon bone. The pasterns and their joints are quite in keeping with the bones above them, and are not so long, straight, and weak, as those of the English horse. The feet are generally in the same proportion; but the Arabs themselves appear to be very careless in their treatment of them. The body or centre-piece of the Arab horse has rarely too great length. This is a very uncommon fault in the pure breed; and there is no breed of horses that are more even in this respect than the Arab. Behind this, we come to a great peculiarity in the breed—his croup: I might say, an Arab horse is known by it: he is so much more beautifully made in his quarters, and in the way his tail is put in, than most other breeds. His loins are good: he is well coupled—his quarters are powerful, and his tail carried high: and this even in castes that have little more than a high bred stallion to recommend them. The straight-dropped hind leg is always a recommendation, and almost all racing Arabs have it; and this, when extended, brings the hind foot under the stirrup, and the propellers being of this shape, give a vast stride, without fear of over-reach. The thighs and hocks are good; the latter very rarely know either kind of spavin or curbs. The joints and processes are preeminently well-adapted for the attachment of the muscles, while the flexor tendons of the hind leg generally correspond with those of the fore. The hocks are not much let down, nor the hind leg so greyhound-like as in the thoroughbred English horse. In stride, too, he is somewhat different, inasmuch as it is a rounder way of going, and is not so extended or so near the ground, but is more like a bound. However, there are exceptions, and I have bred pure Arabs whose stride, for their size, was very extended, and quite like that of English race horses."

In reading this, Captain Shakespeare's admirable description of the Arab horse, I cannot help noticing a sort of workmanlike handling of the subject, which is nowhere to be found in the pages of Daumas, for instance, or in any other of the writings of the—as I must call them—unequestrian nations. In fact, this sort of knowledge, and this style, seem only to be found amongst English, at home and abroad, and would, I am quite sure, be Greek to the generality of foreigners.

In speaking of several of the best breeds of India, the author ascribes their superiority to strains of Arab blood, notwithstanding which the Arab appears to be under a cloud in that country. "With the tide," says Captain Shakespeare, "now turned against Arabs for stallions, at the Cape and in India, I will, nevertheless, prophesy that at no very distant date he will again come into favour."

My purpose in making the above extracts is to present my readers with a view of the Arab in a country where he is to be found working side by side with various other imported horses, and this in a chapter advocating the propriety of breeding cavalry horses in India. Here the writer prefers him to all other horses, and says, that of horses not bred in the country, the Arab and the gulf horse can alone support the climate and work in the sun of India. In his category the Australian horse is particularly mentioned. As a carriage horse and racer he is said to shine, but in no other way. He is held to be exceedingly vicious.

Having said so much of them, there still remains a point to be considered before passing on, and a question to be asked. Are these Arabs of India real Arabs, or is it only an inferior cock-tail Arab that is so immeasurably the first horse in India? There can be no doubt, I believe, on that head; he is not the real Simon Pure. High caste Nedjd horses could be purchased before the mutiny, says Captain Shakespeare, for £70, and now they range as high as £200. About 50 of them, he says, come yearly to Bombay. If a horse could be brought from Nedjd and sold at Bombay for £70, what could he be bought for where he was bred? The price and the abundance of the article both point it out as spurious. Though it is no secret that the Indian Arab is a second class article, I am still glad to bring forward some authorities on the subject; and let the reader recollect that a great many of the Nedjd horses, which though small are of high caste, are sold in Syria. Colonel Churchill, in his Mount Lebanon, says, speaking of Daher, who was shot at Sidon, "The greatest expense he incurred was in blood mares, for some of which he paid nearly £800."

Wellsten, in his travels in Oman, says that the Imaum of Muskat has stallions worth from 1,500 to

2,000 dollars, or £400. Layard, in his "Ruins of Nineveh and Babylon," comes right to the point. He says :—

"The Indian market is chiefly supplied by the Montefik tribes inhabiting the banks of the lower Euphrates; but the purity of their stock has been neglected, in consequence of the great demand, and a Montefik horse is not valued by a true Bedouin. Horse dealers, generally of the mixed Arab tribe of Agayl, pay periodical visits to the Shammar and Aneyza, to purchase colts for exportation to India. They buy horses of high caste, which frequently sell for large sums at Bombay. The dealers pay in the desert from £30 to £150 for colts of two, three, and four years. The Agayles attach less importance to blood than the Bedouins, and provided the horse has points that suit the Indian market, they rarely ask his pedigree. The Arabs, hence, believe that the Europeans know nothing of blood, which, with them, is the first consideration. The horses thus purchased are sent to Bombay by native vessels at a very considerable risk, whole cargoes being lost or thrown overboard during storms ever year. . . . . The Arab horse is more remarkable for its exquisite symmetry and beautiful proportions, united with wonderful powers of endurance, than for extraordinary speed. *I doubt whether any Arab of the best blood has been ever brought to England.* The difficulty of obtaining them is so great that they are scarcely *ever seen beyond the limits of the desert.* . . . . Their average height is from 14 to 14¾, rarely reaching 15 hands. I have only seen one mare that exceeded it. . . . . It is only the mare of the wealthy Bedouin that gets a regular feed of about twelve handfuls of barley, or of rice in the husk, once in the twenty-four hours. . . . The Shammar Bedouins give their horses, particularly when young, large quantities of camel's milk. I have heard of mares eating raw flesh, and dates are frequently mixed with their food by the tribes near the mouth of the Euphrates."

The reader will observe that the first few lines of this extract concerning the sort of horses sent to India involves a contradiction. This I notice frequently! The meaning of the author, however, on the whole, is clear enough, which is, in two words, that the Indian Arabs are not of good pedigree, that a first class Arab has probably never reached even England, and is rarely seen beyond the limits of the desert.

Ten years ago, I met El Signor Angelo, an emissary of the French Government, who had come to Syria to purchase stallions for the Emperor. We parted at Beirout. I heard afterwards that he had succeeded in procuring ten pure stallions for the Emperor, for which he paid £1,000 per head. And after all were

THE ARAB HORSE. 121

they first-class horses? Those who best know the Bedouins say that they never have sold a first-class horse to any unbeliever. On the subject of the difficulty of obtaining first-class Arabs, I will make one more extract, from the many testimonies I have met on the subject. It is from a "Pilgrimage to El Medinah and Meccah," by R. F. Burton, in 1856, who says, " Abbas, the late Pacha (of Egypt), did his best to buy first-rate Arab stallions ; on one occasion he sent a mission to El Medinah for the sole purpose of fetching a rare work on farriery. Yet it is doubtful whether he ever had a first-rate Nedji. A Bedouin, sent to Cairo by one of the chiefs of Nedjd, being shewn by the viceroy's orders over the stables, on being asked his opinion of the blood, replied, greatly to the great man's disgust, that they did not contain a single thoroughbred (Arab). He adds an apology on the part of his laird, for the animals he had brought from Arabia, saying, that neither Sultan nor Shaykh could procure colts of the best strain.

" For none of these horses would the staunch admirer of the long-legged monster called in England a thoroughbred give £20. They are mere rats, short and stunted, ragged and fleshless, with rough coats and slouching walk. But the experienced glance notes at once the snake-like head, ears like reeds, wide and projecting nostrils, large eyes, fiery and soft alternately, broad brow, deep base of the skull, wide chest, crooked tail, limbs padded with muscle, long elastic pasterns; and the animal put out to speed soon displays the wonderful force of blood. In fact, when buying Arabs there are only three things to be considered—blood, blood, and again blood . . . It is said that the Zu Mahommed and the Zu Husayn, sub-families of the Beni Yam, a large tribe living around and north of Sanaa, in Yemen, have a fine large breed; called El Jauf, and the clan El Aulaki rear animals celebrated for swiftness and endurance. The other races are stunted, and the Arabs declare that *the air of Yemen causes a degeneracy in the first generation.* The Bedouins, on the contrary, uphold their superiority, and talk with the greatest contempt of the African horse.

"In India we now depend for Arab blood on the Persian Gulf, and the consequences of monopoly display themselves in an increased price for inferior animals. Our studs are generally believed to be sinks of rupees. The Governments now object, it is said, to rearing, at a great cost, animals distinguished by nothing but ferocity."

When it is borne in mind that all the best breeds of horses, not only of the East, but of the world, all trace their best virtues to Arab descent; and when I recollect the many persons well able to estimate their qualities who have pronounced in their favour, it seems unnecessary, and that there is but little occasion for my adding my humble testimony to that of so many. And yet as a Bedouin would not give a piastre for the opinion of any one, concerning a horse, who was not a Bedouin, so an Australian may not be displeased to hear the opinion of an Australian, and to this end only I will say my say. About ten years ago, I had many opportunities of seeing Arab horses in Syria, Turkey, the Holy Land, and Egypt, and before I saw them I had already had some experience of the horses of England, France, and Spain, besides those of Australia and Tasmania, in none of which countries I had resided less than a year. I had also seen those of Greece, Italy, Flanders, Belgium, Switzerland, Turkey, and other places too numerous to mention, so that I may be said to have approached the examination of the Arab after having seen most of the best breeds in existence. Since then I have had an opportunity of again reviewing them, and if necessary of revising my first impression. In all these countries I have ridden more or less, and had originally in Tasmania and Australia been so unceasingly in the saddle, that it is not to be wondered at that I acquired a habit of glancing my eye over every horse that came in my way, and involuntarily daguerreotyping his figure upon my memory. Neither is it surprising that as I went along I made myself acquainted with any little peculiarities in their treatment, as well as an estimate of their capabilities. At the same time, as this was but the same species of habit which all have more or less in connection with any pursuit that may interest them, I made no notes of my observations, as at that time I had not the remotest idea that I should ever write on the subject, or turn the knowledge thus picked up to any practical purpose.

Of Arab horses, though I have seen many belonging to Pashas and royal personages, to rich men and to wandering Bedouins, I am not sure that I have ever seen one of the most esteemed castes. But, if not, I have seen many that had been purchased at respectable prices, seen many of them at work, and ridden a few of them. A gardener soon forms an opinion of a spade, and a woodcutter of an axe, and so one who has lived in the saddle soon makes up his mind about horses. Mine, at all events, was not long in being satisfied. Instead of seeing anything to object to in the Arab as a saddle-horse—his size excepted—all that I did see, and all that I was enabled to glean concerning him in his native land, only led me the more decidedly to endorse the opinions of those numberless very competent judges who had gone before me. I never met a man who had tried him and did not like him. I found him in speed inferior to the horse of England, but in tractability, constitution, durability, soundness, abstemiousness, temper, courage, and instinct, eclipsing and surpassing all other horses that it has been my chance to meet with. In that quality so pleasant to the horseman, sagacity, I think he has no equal. Even half-breeds sprung from him are remarkable in this point. When in Syria, I bought a little horse which was no beauty, but had evidently got a good deal of Arab blood in him. He stood 14.2, and was six years off. I had him about three months, and rode him, perhaps, a thousand miles. He was never fed more than twice a day under any circumstances. At sunrise his breakfast, which consisted of two double-handfuls of barley, was given to him in his nosebag. This took him, as I often noticed, about twenty minutes to eat. Two hours after this he was allowed about four quarts of water. Three hours before sundown he was taken to water and allowed to drink his fill. Two hours later, ten double-handfuls of barley, sometimes mixed with three or four handfuls of chaff, were given to him. Such were the habits in which he had been brought up, and such the amount of food which proved in every way sufficient for him even when at work. Once on my journey, at Tyre, I took it into my head to alter all this, and after a long day to give him, when I pulled up, as much water as he chose to drink, followed by his barley, without allowing any interval to elapse. The result, as might be expected, was, that next day he was unwell. However, a day's fast, and a return to his old regimen, set all right again. On this he looked as round as if he had eaten as much as our Australian horses are accustomed to do. For 20 days I rode him 30 miles a day. He had fifteen stone on his back—the country was mountainous and rocky, but he never made a false step, and I think he improved in condition. I never remember to have found him weary.

His intelligence was quite beyond any single instance I have ever witnessed in an Australian horse. To say that he recognised his master as one man does another, would hardly be doing justice to his sagacity: he rather seemed to recognise me as the detective does his man. If, as was sometimes the case, he was in a stable with a hundred others, where many persons would be constantly passing to and fro at all hours, he seemed to be constantly on the watch for me. My voice, of course, he knew at once. The sound of my footsteps seemed as familiar to his ear as was my appearance to his eye. He would greet me with his voice when I was several hundred yards from my tent.

I might adduce a thousand instances of instinct, such as one would rather look for in a dog than a horse—but as this is apart from my subject, I will spare the reader my reminiscences.

It has commonly been supposed of late years that

the difficulty of acquiring first-class Arab horses and mares has been exaggerated, and, notwithstanding the authorities I have adduced on that head, I feel that there may be a lurking incredulousness, which perhaps actual contact with the Bedouin can alone dispel. But, besides his fondness for his mare, his customs, religion, and self-interest are all against the sale of his best horse. To ask a Bedouin, who has not volunteered to sell, to part with his horse, is no less an insult than it would be to ask an English nobleman, under the same circumstances, to put a price on the halls of his ancestors. "It is a great offence," says Daumas, "to say to an Arab, 'Will you sell your horse?' before he has himself made known his desire to sell: 'I am thought to be in great misery,' says he, "when you dare to make me such a proposal.'" But a still greater difficulty in the way is, that to no one is a good horse so necessary as to the Bedouin. "*L'Arabe*," says Daumas, "*ne peut mener que la vie à deux, son cheval et lui.*" Layard puts this in a clear light, where he says: "To understand how a man, who has perhaps not even bread to feed himself and his children, can withstand the temptation of such large sums, it must be remembered that, besides the affection proverbially felt by the Bedouin for his mare, which might perhaps be not proof against such a test, he is entirely dependant on her for his happiness, his glory,

and, indeed, his very existence. An Arab, possessing a horse unrivalled in speed and endurance, and it would only be for such that prices like those I have mentioned would be offered, is entirely his own master, and can defy the world: once on its back, no one can catch him. He may rob, plunder, fight, and go to and fro as he lists. He believes in the word of his Prophet: "That noble and fierce horses are true riches." Without his mare, money would be of no value to him: it would either become the prey of some one more powerful and better mounted than himself, would be spent in festivities, or be distributed amongst his kinsman. . . . . .
No one has a heartier hatred of restraint than the true Bedouin: give him the desert, his mare, and his spear, and he will not envy the wealth and power of the greatest of the earth. He plunders and robs for the mere pleasure and excitement which danger and glory afford."

Such is the Arab horse, and such are the ways of his master. I do not know where you will have to go to find such another! Such he is now, and if we may trust the accounts of old travellers, such he has long been. In the "Voyages de Monsieur Le Chevalier Chardin, en Perse, 1709," we have a description of him, which singularly tallies with that of General Daumas: "J'ai dit," says he, speaking of the Persian horses, "qu' ils sont *les plus beaux de l'Orient ; mais pour cela ils ne sont* pas les meilleurs, ni les plus recherchez. Ceux d'Arabie les passent et sont fort estimez en Perse à cause de leur légerté, car ils sont, quant à la forme, semblables à de vrayes Rosses par leur taille séche et décharnée."

But though mentioned last, not the least item of the glory of the Arab still remains to be spoken : the renown of his descendants in foreign lands! Where is there a horse excelling those around him who does not trace his virtues to some strain of Eastern blood? To him England owes all her equine glory; to him are indebted for their best blood—North America and Australia, France, Spain, Italy, Russia, Circassia, and India.

Such is the Arab horse, who, as Youatt says of the Barb, "Presents the true combination of speed and bottom ;" and whilst we remember that he wants but two inches in height to be the perfection of horseflesh (which want, I firmly believe, more plentiful food would radically supply in three generations), let us not forget his renown as a sire, his sure-footedness, docility, beauty, speed, abstemiousness, stoutness, and courage—where shall we find his peer? All honour to the little horse!

Who will contradict me when I assert, that no breed of saddle-horses has shone, which has not possessed some strain of Arab blood? Who will show me that I am incorrect, when I say, that the virtues of European breeds are in exact relation to their affinity to the Arab?

Who that has known the Arab has not preferred him to all other horses?

THE ANDALUSIAN HORSE.

"Yo no quiero criados que tengan un aspecto tan
virtuoso, porque estoy escarmentado de ellos."
Gil Blas.

Andalusia was long celebrated throughout Europe for its excellent breed of horses. From that kingdom the Spanish knight and soldier got his charger, and the Hidalgo his jennet. After a considerable but gradual falling off in the quality, and deterioration in the figure of these horses, the studs whence they were drawn were seized for cavalry remounts, and entirely broken up by Soult and other French marshals in the great Peninsular war.

The best blood of the Andalusian horse was that of the Barb, imported by the Moors, and from it our thoroughbreds are partly descended. Some vessels also of the celebrated Armada, with Barb horses on board, were shipwrecked on the coast of Scotland and the Scottish Isles, whence sprung the

Galloway, and several other breeds of cobs and ponies, which long showed traces of their Eastern blood, and were justly celebrated for speed, bottom, and beauty.

Of the Spanish horse, but little is said, either by Youatt or Stonehenge. The latter author attributes great courage to him, which he is said to evince at the Bull fight, meeting his horned antagonist without flinching. Flinch, indeed, the poor living skeleton who is condemned to the *Corrida* does not : whether this be from courage, or that he feels his life to be too miserable to be worth the preserving, I know not; but at all events his off eye is always bandaged, so that he cannot see his antagonist, who is always received on that side, which may in part account for his valour. From two to ten of these miserable scarecrows, according to the state of the funds available for the purpose, are served out to each bull, and are slaughtered by him in the most cold-blooded, unresisting manner, and forms, in fact, the disgusting part of the exhibition.

No one catches the eye sooner, or looks more gay, showy, or attractive, on his horse, than the young caballero of Andalusia. In all the great inland cities of that province, you still see him take his daily ride in his national costume; he still reins back his steed with his long Eastern curb; *el paso* is still his chosen gait. The swarthy complexion of the rider, the stately tread of his horse, his Moorish saddle and shovel-stirrups; the gorgeous shawl that falls across his knees, the many coloured housings, and horse-tail tied up with ribands, irresistibly recall to memory, as he passes down the narrow, cartless streets of Calle Francos or el Zacatin, the days of his Moorish predecessors: the stormy days, when Bobadil el Chico sallied out of the gate of Elvira to avenge the fall of Alhama, at the head of all the chosen chivalry of Granada, mounted on their celebrated bay mares.

> "Por esa puerta de Granada
> Sale muy gran Cabalgada
> Cuanto del hidalgo moro,
> Cuanto dela gegua baya."

With his love of display, something too of the Mahometan love for his horse has been bequeathed to his conqueror, whom some one has termed "an Arab in a house," and, as in the desert he shares the tent of the Arab, so in Andalusia he is lodged in the house of his master. The houses of Andalusia are usually quadrangular buildings, with no back yard of any sort, and with but one entrance, which is through a large iron or bronze gate, fronting the street, in the centre of the façade, and at the entrance of the spacious hall, on each side of which is a sort of anteroom, where the horses are stabled:

hence the stranger is surprised when he sees, for the first time, the saddle-horse, not led out of a stable, but out of the front door of some palatial edifice. A few steps onwards takes you into the *Patio* or court, which is the centre of the house, and forms the summer sitting-room. The external edges of the Patio are overhung to about the depth of ten feet all round, by the projecting balconies of the first story, and the rest of it is shaded in summer from the sun by a rich awning. It is a very pleasant place in the hot weather! A sparkling fountain in its centre—another legacy from the Moors—drops with silvery sound into a marble font, cooling and bedewing the air, and the tesselated floor, on which stand here and there statues, and orange-trees, flowers, and green shrubs, growing in pots. Under cover of the overhanging balconies are paintings, tables, piano, and guitars; books, and work-boxes, rich carpets, and sofas, on which, as you walk by, you may see seated the dark-eyed ladies of the house, under the same roof, and but a few yards from the family steed, whose neigh is often heard to answer the voice of his master. The bath is close at hand; and many a gay lamp lights the scene at night.

> " Sostenian los ricos arquitraves
> De sus claros moris-cos corrideres
> Columnas ligerisimas. Sus naves,
> Adornaban arábigas labores,

THE ANDALUSIAN HORSE. 135

Sutiles cual la pluma de las aves
Tan brillantes como ella en sus colores ;
Frutales desde el huerto à las ventanas
Alargando limones y manzanas.

" Sus patios, que en albercas espaciosas
Reciben unas aguas cristalinas
Al cuerpo gratas y al beber sabrosas
Pilas eran de bano alabastrinas
Sembrado el borde de arrayan y rosas,
Donde las bellas moras granadinas
El seco ardor de la mitad del ano
Ahuyentaban di si con fresco bano."

In bringing before my readers the horse of Andalusia, I have a very special purpose. It is not, as may be imagined, to point out any particularly good qualities that he may possess—and he is not without several, for in good specimens he is still healthy, vigorous, and a weight carrier, abstemious, and but little subject to disease, and, in fact, still before many of the other horses of Europe—nor in any way to hold him up to admiration. It is rather to speculate on his decline and debasement, and how it came about; how he is now so altered from what he was when the last King of Granada led back his Moors to Africa. When I first landed in Spain I was surprised to see so little of the Arab figure about the horses ridden by the gentry, and to find that any of it that still existed was to be met amongst those but little esteemed, and put to the meanest drudgery. In talking with Spaniards on the subject, I soon discovered that their taste in horseflesh was quite at variance with an Englishman's or an Australian's ideas on that subject; and that the form and action of the thoroughbred, whose speed they made no difficulty in allowing, were both equally disliked in Spain. I had at first some difficulty in believing that the opinions which I heard expressed could be those of connoisseurs. That such was the case, however, I soon had an undeniable opportunity of learning. This happened on the occasion of a great show of saddle-horse sires, at Seville, which took place in the *Corrida de Toros*. Curiosity led me that way, and being in the boxes I had a good opportunity of seeing the horses as they were led round the circus. Being in company with a gentleman who knew one of the judges, who stood in the arena, I got introduced to him, and through his kindness was allowed to descend and inspect the horses with the judges. Here I heard all their points discussed, and heard our old adage that "like begets like," which Cervantes expresses, "*cada cosa engendra su semejante*," pretty freely made use of, and saw the prize awarded. The whole affair amused me a good deal, and I recollect the discussions on the subject well, though I made no memoranda of the matter, as I had not at that time, nor till long after, the remotest idea of ever writing a line about the Andalusian or any other horse. Here I found the great contrariety of their opinions and those of Englishmen; that what we prize they often dislike, what is valued by us is often held cheap by them. Thus, to our light and somewhat lengthy neck, they prefer one short, high-crested, and thick; a lean, fleshless head, and an open, intelligent eye, are not in the category of their beauties. We esteem a light, firm, and active walk, they the *paso*. The shape of bone, whether flat or round, cleanness of sinew, and proper formation of joints, are little thought of. They are much less observant of points than we are, being rather inclined to restrict their criticism to the carriage, action, and *tout ensemble* of the appearance of the animal; not deducing from experience, that certain make and forms are excellent for certain purposes, but as embodying their idea of a sort of pompous or parade beauty, arbitrarily agreed upon, and fixed by fashion amongst them, as being the beau-ideal of perfection. If the horse is high, say from $15\frac{1}{2}$ to 16 hands, a good colour, bay being preferred as of old, luxuriant mane and tail, short back and neck, body rather round than deep, wide chest, with strong arm and gaskin, good sound feet, wide hips and muscular coupling, performs the *paso* to their satisfaction, carries heaps of flesh and a good coat, is quiet, but proud and fiery in his gait, he quite comes up to their idea of what a horse should be. Something, perhaps, of the figure that a Suffolk-punch mare might throw to an Arab sire. His shoulder straight and heavy, his neck short and thick, his pastern straight, his bone round, his leg fleshy, and his head coarse, will not be seriously objected to, or in fact noticed.

He is ridden with a strong heavy curb, which cannot be denied, his head is held high, and his chin forced close to the lower part of his neck. Held in this posture, a long spur applied to his sides, produces the gait known as *el paso*, in which the horse seems to walk with his hind legs, and lift his fore feet as if at a jog-trot, but very high and with a very awkward turn of the foot, throwing up the dust and mud; but being restrained by the bit, he plants it quickly and with a blow, but very little in advance of where he raised it from. This action of the fore feet is much admired by the initiated. Here, then, were before me the fact and its explanation. What sort of horse is admired in Andalusia, and how he has been brought to his present shape and action from what he once was. The metamorphosis presents but small difficulty in the solution. Once fix on a certain configuration as a beau-ideal, horses of that figure become saleable; the breeder anxiously seeks for sires embodying the desired characteristics; stallions of that stamp, rare perhaps at first, become more plentiful from selection, as the offspring of those that exist, and the figure sought at length becomes general. I am well aware that there are limits set by nature to this principle, but, as I am treating the subject practically and not scientifically, there appears to me no occasion for entering on the discussion of that point. Neither must my reader imagine that I mean to state that the present Andalusian horse has been produced without any crossing, Crossing with base blood, is, of course, the high road to deterioration, and has, no doubt, facilitated the change effected here. I will, however, add, that I have no doubt but that by a skilful use and selection of climate, food, and sires, the little Shetland pony, might, after perhaps a hundred or two generations, without any cross, be brought to rival the size of our large brewer's horses, or that mighty animal be dwarfed down to the diminutive height of the Shetlander, or to almost any existing intermediate configuration, just as food, climate, and customs, have given us, in our own race, the pigmy Astec and the gigantic Patagonian.

So much for the saddle-horse of Andalusia. The more humble nag of the *caleso*, stage coach, &c., whilst retaining more of the figure of his Moorish ancestor, has been reduced by hardships and neglect to a very contemptible condition. Still, however, I have noticed that he is more spirited, more abstemious, and more stout, than a horse of equal figure would be in England, Belgium, or France.

Encouraged by the taste of the English garrison at Gibraltar, a few somewhat Arab-like looking horses are bred in that neighbourhood for sale, and as they have been most brought under the observation of Englishmen, their figure has often been supposed to be that usual in Spain. Such, however, as I have said, is not the case, for the old Arab blood and figure may be now said to have been perfectly and systematically debased in that country, and a very second-rate animal to have occupied his place, offering another proof, if it were needed, of how completely a wrong aim in breeding may ruin the best race of horses.

THE
AUSTRALIAN SADDLE-HORSE.

"Bid the ostler bring my gelding out of the stable."— *Gadshill.*

We now come to the horses of our own country. To Australia, as to America, the horse was not indigenous. He was first imported into New South Wales from England. Mares were subsequently brought from the Cape of Good Hope, from Valparaiso, and pony mares from Lombock and Timor, and have now increased to very large numbers. Thoroughbred sires have been imported from England in constant succession, and in considerable numbers, some few of them being of good quality, the rest mere weeds, which have only served to deteriorate our breed. Arab sires, with one exception of low caste, have also frequently been introduced from India and the Cape, but not in numbers sufficient to have had much effect on the breed here. Such as have come have been chiefly Gulf Arabs of the racing-type ; greys, of little bone and low caste.

In the early days of New South Wales considerable care was used in the breeding of saddle-horses ; sires were selected with some judgment, and fillies allowed to arrive at a fit age before they were sent to the horse. It also happened that several of the stallions of the early days proved themselves to be much more fitted to be the sires of saddle-horses than any which have been since imported. This I have heard remarked by many competent judges, and have myself witnessed to a certain extent. Amongst the most fortunate of our early sires were Whisker, a thoroughbred, and Satellite, an Arab, in New South Wales ; and Peter Finn, a thoroughbred, in Tasmania. These horses, compared with what we have since had, got excellent stock. There were also one or two good stallions in Sydney still earlier than those just mentioned, but their names I am no longer able to recollect, nor do I know that it is of much importance. The horse has increased rapidly since his first introduction to Australia, and is now found in large numbers (in many places neglected, wild, and unclaimed) in all the settled districts. Though, as might be expected, in countries so dissimilar in position, climate, and pasture, as those now settled in New Holland, some peculiarities in his qualities have begun to manifest themselves in relation to these causes, and certain distinct characteristics to spring up in various localities, yet on the whole, a great similarity still exists in all, management everywhere having been much alike, and on these points I will first speak.

On the first importation of horses into New South Wales, the animal was too valuable to be neglected, and the demand being limited held out no inducement to owners to sacrifice quality to quantity by too early breeding. As the animal increased in numbers, a degree of stoutness and capacity for work quite unknown in England began to manifest itself in the colonial-bred horses, and the English emigrant was surprised to find that the horse he had bred in the country of his adoption was remarkable for a vigorous health and freedom from sickness and disease, and would endure fatigues and perform journeys on grass-feed alone, which would have overtaxed the parent stock in England on the best stable keep. This robust health he still possesses in a remarkable degree. His stoutness was still further increased by the importation of thoroughbred stallions, who, be their pedigree what it might, have had no equals as foal-getters amongst those since introduced. The racing and Arab blood of Whisker and Satellite seemed at their meeting to flow with fresh vigour in the land of the South. As our horses increased in number they still continued not only to uphold, but even to add to their early reputation, thus substantiating their first testimony as to the fitness of the Australian pasture and climate for their full development and perfection. The gentry now, both in town and the bush, rode horses of rare stamp, whose performance surpassed even what their figures promised. The Mounted Police and Border Police, heavy men frequently, with heavy accoutrements, bestrode chargers such as no horse-regiment in England could or can turn out, and have been frequently known, as I have been informed by credible persons, to do their 70 miles for four or five consecutive days without difficulty. Of course I mean on grass-feed alone, without hay or corn. To this day, it may as well be said, after leaving town and passing through what may be called the country districts, the horseman enters the bush, where artificial forage of any sort may be said to be unknown. Hence, when speaking of riding in the bush, it must be understood that the horse gets nothing to eat but what he can pick up, which is often but a very scanty allowance.

A reputation at last begun to be established by the horses of New South Wales, or, as it was then better known as, Botany Bay, and they began to be exported for the use of the British cavalry in India. These I have been told by the late Sir George Arthur, Governor of Madras or Bombay, I forget which, and other officers, fully kept up in India the character they had brought with them from Australia, both on the road and on the turf; and being weight-carriers, large and well-bred, bringing to mind the chargers of the old country, were preferred to all others at that time in India for that purpose. But these days of early glory were not destined to be of long continuance. Colonization began to progress, and the town or farm-reared Englishman began to learn a little bush craft. He had already burst through the Blue Mountain barrier, which had so long held him back. The sheep and cattle originally imported had increased rapidly, and fresh pasture had become necessary. The services of the horse began now to be more requisite, and he had fallen into the hands of breeders, who, however they might excel as horsemen, knew little of the rules necessary to his production, or the preservation of his good qualities. As his position became remote from civilization, the services of imported, thoroughbred, or picked sires, became unattainable and expensive, and colts often of the most indifferent descriptions were substituted in their stead, and ran loose with the little mobs of mares and fillies which began to be found about the homesteads of the settlers; fillies of from two years to thirty months old began regularly to rear their foals, and the horse stock, as a consequence, rapidly

H

to deteriorate. Then fresh settlers and more capital began to find their way into the colony, and Victoria (then Port Philip), and Adelaide, new settlements, offered good investments to the monied, and the excitement of emigration and adventure to the young stock-holder. Horses that at this day would not realize over fifty shillings, were then eagerly caught up in these new markets at £70, £80, and £100, and their production as a consequence was at once pushed to the utmost.

This brings us to the year 1840, when the author first landed in Melbourne. Horses were then coming slowly from the Sydney country to Victoria, generally poor, stunted, miserable wretches, the culls of that district, whose degeneracy weedy race-horses had begun, and early breeding and haphazard sires had completed. But, in those days, from time to time came into Victoria an occasional veteran Satellite horse, sinewy, of exquisite make, and of great utility and power; the remnants of the original horse, the had-been of other days. The young stock had become leggy, girthless, boneless abortions, that had dwarfed down to from fourteen hands to fourteen hands three inches, whose only value was their market price, for intrinsically they had none. Even this value soon disappeared, for before 1843 had passed away, three of the former £60 horses had become equivalent in the market of New South

Wales to the price of one saddle. In fact they became almost valueless.

When this commercial crisis had pased away, and the wool and tallow brought into the market had again made the country solvent, stock of all sorts reaped the benefit of improved management, with every other interest. The horse amongst the rest. The hack now rose in value, from a mere song to about £12. People, however, began to find that they had got into a bad sort, that the little weeds now in use were unable to carry them, and the old settlers and Sydney natives looked back with regret to the old days when they rode those fast, up-standing, strong, and fiery horses, far inferior, it is true, to the Arab, yet whose like were now nowhere to be obtained. More substance ! big horses ! began to be the cry. Away with these woolly-legged wretches ! Let us have the tall bony horse of the old day ! Breeders prepared to supply the demand. The weedy, delicate, washy, unsound, imported thoroughbred began to be put to the heavy cart mare. The produce of his cross was to be the sire for the future, and great results were expected. This sire showed an increase in size, his bone measured more than did that of his sire, he had some go, but still he failed to satisfy the public. For "no breed can be improved," says Captain Shakspeare, "by so great a disparity in the sexes; the produce will be entirely misshapen, and their bodies and limbs out of all proportion." The broken kettle had been patched with wood and not with iron. The sire himself proved but a poor saddle-horse, his get was ill-descended on both sides, and of the wrong sort. How could it be otherwise? This large bastard sire was put to undersized, unroomy, mares. Sometimes the result was a heavy carcase and insufficient bone; at other times a light carcase and soft unwieldy legs, with narrow chest and short shoulder, small girth, and delicate constitution. Each a disappointment, both bad! What was termed the substance of the sire, and the speed of the weedy dam, did not manifest themselves. Fingers had been fixed on the table, but it would not turn! Spirits had been called from the "vasty deep," but they only whistled and looked another way! Whatever might have been effected, *which is not much*, by more gradual crosses, had been entirely frustrated by the extreme and sudden measures of the breeder. He had forgotten that if the Australian is always in a hurry, nature keeps on at her old established pace; she improvises nothing! With her the superstructure is always in relation to its base; perhaps he did not know that the tree centuries old may be uprooted, cities rise on its ruins and become magnificent and powerful, and, I had almost said, decay again, in less time than nature has fixed as the possible period in which a debased breed of horses can be reared to the acme of perfection. For, whatever is to be real and effectual in this matter, must be gradual. The oak that is to endure for ages, must not be forced like the mushroom!

The cross turned out a failure, and some breeders imported, in place of weedy racers, thoroughbreds of more bone and substance. But where do we notice the anticipated results? Besides, our horse-stock had become so numerous, that to provide good sires in numbers requisite to re-establish our breed was impossible.

In the meantime, the great English specific for the improvement and perfection of saddle-horses, namely, races, had not been neglected. The purchase of thoroughbreds in England for this purpose had been frequent, and the animals imported of considerable pretensions. Jockey clubs in Sydney, Melbourne, and Adelaide, offered prizes for excellence with no niggard hand; the betting rings were well attended, betting brisk, and black legs, &c., I believe became acclimatized and flourished. Races, too, at the country towns were frequent and well attended, and the breeders of thoroughbreds numerous. In fact, I know no country where racing is, and has been, carried on so extensively in proportion to its population as in New South Wales; none where it forms so general a subject of conversation and of interest. In that district you cannot find a township where there are but half a dozen huts congregated together that does not boast its annual races; hardly a roadside bush-public-house that has not its rude course. I have seen races run over stony ground, hilly ground, hard plains rent with wide fissures, and over country so heavily timbered—where no better was to be had—that many a new arrival in the country would have found it unpleasantly close to ride to hounds through. The *time* of our races has always I believe been respectable, and certainly the speed of our racer has been acknowledged in India, the only country to which he has been exported.

If such has been the system of our horse-breeding for the last forty years, it will now be possible to say something definite respecting its results. We have had racers and Arabs of a certain class, and hunting sires; we have seen crossing and re-crossing in every imaginable way; a thousand breeders have tried every blood, from that of the diminutive Timor to the gigantic Clydesdale, in many different proportions. Each favorite nostrum has been fairly tried, and with what result? A constant and progressive degradation of our stock. Has anything been left untried? How comes it that whilst breeding a pure racer for the course, it has never occurred to us to breed a *pure saddle-horse* for the road or the bush, as the Saharian or Bedouin does?

As regards the racer, it may be stated, that his condition has not undergone any decided change, either for good or bad, since his introduction here; neither has any particular stud or stable been remarkably in the ascendant. If any race-horse has attained pre-eminent celebrity in Australia, it was, perhaps, old Jorrocks in his day, whose stoutness and long lasting qualities, and the number of years over which his winnings extended, joined to a fair degree of speed, are not unworthy of notice, and I insert for the benefit of my readers an account of him from *Bell's Life in Sydney.*

"Near the sequestered township of Mudgee, in New South Wales, in the October of '33, Matilda dropped a bay horse foal to Whisker, both sire and dam being then the property of Mr. Henry Baley, of Baley Park, an influential and highly popular magistrate of the territory, who, in conjunction with other members of his family, for a series of years devoted large pecuniary resources, and multiplied experience, to the improvement of the breed of colonial horse stock. Their success was happily as eminent as their endeavours were philanthropic.

"The pedigree of this quadrupedal phenomenon, whose extraordinary racing powers, like those of the Eclipse, so long remained in embryo, may thus be quoted:—Jorrocks by Whisker, out of Matilda, by Steeltrap; grandam Vesta, by the Arabian Model; great grandam Caraboo, by Old Hector; great great grandam by Rockingham, etc. Whisker (imported by Mr. Henry Baley) foaled in 1829, was got by Whisker out of Woodbine, sister to Fortuna by Comus; grandam by Patriot; great grandam by Phenomenon out of Czarina by Babraham—Blank, etc. etc.

"Matilda once occupied a prominent position on the local turf, and Vesta, bred by the late Nicholas Baley, Esq., proved herself staunch in a severe race of two-mile heats, snatching the laurel from the following high-pressure nags, viz. :—Mr. W. C. Wentworth's Currency Lad, by Stride; Sir John Jamieson's Abdallah, by the Arab of that name; Mr. Icely's Alraschid, by ditto; and Mr. Lawson's High-flyer, by Old Hector. The mare took the sweep, after a terrific struggle of four heats.

"Jorrocks is a bright bay, with black points, stands fourteen hands two inches, shows rather a short head with a broad forehead, full and beautiful eye, a somewhat close nostril, handsome racing neck, slightly swayed, fine oblique shoulder with great depth, wide powerful chest, good muscular forearm, rather long knee, short thence to the fetlock, by no means a long pastern, round handsome barrel, very deep girth, short back, which dips near the setting on of the couples—an extraordinary depth of rib —very broad across the loins—a tail well set on— powerful buttocks—strait hocks—round feet, and remains perfectly free from sparvin, splint, curb, or corn. His *tout ensemble* displays the lengthy sloping angular shapes of the English blood-horse, with much of the wiry elasticity of the desert born. Honest, courageous, and highly tractable, as he ever proved, we cannot help regretting that the colony should have been deprived of his valuable services for stud purposes; but so it is, and regret is useless.

"Thus, having as freely as seems necessary alluded to his pedigree and appearance, his performances seem to be the next thing that demand retrospection.

"What induced the owner of Jorrocks to keep him so long out of work does not appear; but it is certain that till the close of the year 1836, he was allowed the parole, undisturbed by the chamion of the snaffle, and unruffled by either the whip or the currycomb. About this period he was broken in, and kept upon easy duty till he had well-nigh completed his fifth summer, when he was sent to a cattle station on the river Barwan, and in the humble drudgery of a stock-horse and hack alternately, he continued for the four following years. During this pilgrimage, in occasional sallies after the bounding denizen of the Australian forest, his untiring speed became apparent ; for, though never stabled, rarely groomed, and almost always exposed to that unenviable concatenation of anti-conditionals, of which our ever fluctuating seasons are productive, to wit absence of pasture and of water, made more agreeably conducive to health by burning winds, brobdignag mosquitoes, and bush fires, though not perhaps absolutely suffering from this quintette of grievances, the gallant son of Whisker, with twelve stone on his back, could frequently turn a flyer.

"Thus encouraged to the belief that the horse knew how to put his 'best foot foremost,' Mr. Charles Baley entered him in a sweepstakes of twenty-five sovereigns each, h ft, which he carried away from four less fleet competitors. The race came off on Coola Plains ; but no notice has been taken of the event by persons who have hitherto endeavoured to trace his performances. Stronger proof of the racing capabilities of the debutant was deemed necessary, and arrangements were accordingly made for the purpose of testing to a nicety the extent of his pace. He was placed under *surveillance* of Old Brown, the *fidus Achates* whose graven image considerably adds to the interest of our sketch. Brown then had Helena and others in training at the stables of a gentleman near Windsor,

THE AUSTRALIAN SADDLE-HORSE. 155

so that abundant opportunities offered for affording trials either against time or blood. That the result of these gallops was in accordance with favourable surmises, may be gathered from the fact of Mr. Rouse's anxiety to possess the horse—a desire which he speedily succeeded in accomplishing, as Mr. Baley, in the commencement of '41, was induced to part with Jorrocks, taking in exchange for his valuable gelding eight springing heifers, equivalent to about the sum of forty pounds sterling. No time was now lost in preparing the horse for the ensuing Homebush Meeting, where on the 16th of the succeeding March, he was brought to the post for the Ladies' Purse of £50, open to all horses carrying eleven stone—three miles. Chesnut Prince, Slasher, Creeping Jane, Tom Thumb, and Paddy were opposed to him—running a good second, he was defeated by the Chesnut.

" On the 30th of the same month he was found at Bathurst contending with Kate Nickleby and Eleanor for the Town Plate of 100 sovereigns, where, racing under the mysterious name of Jollop, he met with a similar physicking. The following day, however, hailed him the winner of the Town Plate of £50, though opposed to Theorem and the Nickleby mare. Theorem broke down whilst running up the distance, and Kitty bolted from the start. On the same day he ran second for the Ladies' Purse, which was taken by Skeleton; Jorrocks being second in both heats, Lushington, Glaucus, and the bolting mare being also in the struggle; Miss Kate served her Jock as in the preceding race. At Parramatta on the 27th April he won the Town Plate of 100 sovereigns, beating Flirt, Prince, Eucalyptus, and Stranger. 'Jollox will be a troublesome nag to beat at long distance in heavy ground,' was the astute remark committed to the broad sheet by the prophetic reporter of the day. Returning to the Hawkesbury, he bore away in triumph a plate of 100 sovereigns, but defeat was early in attendance, as at Homebush on the 24th of August the Australian Stakes were easily taken from him, Plutus, and Eucalyptus, by Mr. Charles Smith's Beeswing; the Ladies' Purse of £50 on the 26th set him again a-foot, but Mentor proved more than his match; yet the Beaten handicap served somewhat to console his owner to the disaffection of the stables of Gohanna, Prince, Plutus, and Planet. His next appearance was at the Parramatta Spring Meeting, September 14th, where he achieved a victory over Plutus, Tory, Planet, and Gohanna, for the Union Purse of 100 sovs., and satisfactorily closing his career of '41 by walking over for the Members' Purse of 50 sovs. on the 14th of the same month.

"The campaign of '42 opened brilliantly for the

old horse, at Homebush, for the Metropolitan Stakes of 75 sovereigns. The Quail, Tranby, Planet, and the Marquis, had to succumb to him, and two days after this signal and easy conquest he also rewarded Mr. Rouse with the Cumberland Cup of 100 sovereigns value, added to a sweepstakes of 10 sovereigns each. Ten horses had been entered, but the winner, Eclipse, and Eucalyptus, only came to the post. This splendid and massive vase was manufactured in Sydney by Messrs. Cohen, of George-street. Then followed a rapid succession of good fortune—to wit: The Mudgee Plate of 50 sovereigns, on the 17th May—the Settlers' Purse of 50 sovereigns, on the 20th, taken from Grimaldi, and the Inkeepers' Purse of 40 sovereigns, on the same day—the Hawkesbury Turf Club Purse of 100 sovereigns, on the 27th of July, when he ran a dead heat with Beeswing for the first heat; but eventually took the race with ease. At starting, lots of money was laid at five to three on Beeswing. Allowing one day to intervene, he again did the needful for his lucky proprietor, taking the Richmond Purse of seventy-five sovereigns, with a sweep of five sovs. added, although an extra 7lb. had been imposed. Beeswing, Quail, Jerry Sneak, and the Marquis were in the affair, but the first named broke down in the second heat and was much injured; the Marquis also fell lame while running a good second to Jorrocks. Early in August, he was defeated by Sir Charles at Parramatta for the Union Purse, and on the 7th, at the same meeting, the Quail deprived him of the Town Plate. The severity of his work at Windsor, but a few days previously, accounts in some measure for the want of temporary success—'the old horse had evidently had much of it lately, and his condition being decidedly bad,' so wrote the sporting chronicler of the day. He was next entered at the Champagne Stakes at Homebush, in September, but was drawn to let Quail walk over. From this period a considerable spell was allowed him, not having been called on—

"To gambol o'er the grassy floor"

till the 11th of April, 1843, when he was beaten by Eucalyptus, for the Members' Purse of 100 sovs. The Beaten Purse was all that fell to his lot. At the meeting at Homebush, on the 22nd of May, his star became once more ascendant. The Metropolitan Stakes of 100 sovereigns, falling to his stable, in defiance of the efforts of Sir Charles, Tranby, Blacklock, and Eucalyptus, and on the day but one following he took the Cumberland Cup, of 100 sovs., from Blacklock and the Quail. At Parramatta, in the following July, he experienced another reverse, being beaten by the Marchioness, Harry Lorrequer, and Œdipus, for the Union Stakes, but it was thought by many, that had his jockey, Higgerson, made play earlier in the race, he would have had a much better place.

"The Town Plate of 70 sovs., with a sweepstakes of five sovs. (a single event of three miles seven hundred and fifty yards, run in 6 min. 58 sec.), July 14, tended to resuscitate his fame, to the disparagement of Lorrequer and Quail. On the 12th of September the veteran visited his favourite arena at Homebush, where he triumphed over Lottery, Marchioness, and Harry Lorrequer for the Champion Cup, Lottery running second in the three mile gallop. The Champagne Stakes of this year were also awarded him. In the November following he honoured Penrith with his presence, and, after a dead heat with Sir Charles, took the Town Plate easily. The Baronet was subsequently more fortunate, and snatched the Publicans' Purse. (Jorrocks was now running under the ownership of Higgerson.) Thus finished his exploits of the year ; nor did his name adorn the Calendar till the middle of April '44, when at Windsor he was thrashed by Harry Lorrequer in the Town Plate Race on the 10th, but took the Richmond Purse of sixty sovs., on the 11th, Lorrequer and Election also being candidates for the emolument. At Homebush, July 23rd, he ran second to Lottery for the Metropolitan

Stakes, when the three miles was done in 5 min. 50 sec., Jorrocks losing by half a length; but on the 25th the tables were turned, Lottery playing second fiddle to the Old'un for the Cumberland Stakes of fifty sovereigns. He next was to have been met with at Berrima, where none dared oppose him, and the easy task of cantering in solitary grandeur round the heath secured him the Town Plate of forty sovereigns. Hence he proceeded to Goulburn, where his determined antagonist Lottery was in waiting to draw a prize of £40—viz. the Innkeepers' Purse; heats, two and a half miles. They next met at Penrith at the close of October, where Jorrocks again drew blanks, Lottery winning the Town Plate on one day, and Election beating him on the next for the Publicans' Purse. The Beaten Handicap flung a few crumbs into Higgerson's hat.

"Under somewhat brighter auspices the season of '45 opened at Homebush May Meeting, where he licked Lottery in a three-mile race for the Metropolitan Stakes of sixty sovereigns; Lottery went on the wrong side of a post, supposed to have been so contrived intentionally. But in return Jorrocks was beaten by Blue Bonnet for the Bungarrabee Purse. Heavy sums exchanged hands, the odds having ranged as high as five to two, and in some instances three to one, the Old'un against the field.

In August he was again beaten by the mare at Maitland, where he appeared as the property of Mr. James Doyle. The annual account was closed at this meeting.

"In April, 1846, he took the Hawkesbury Town Plate of fifty sovs.; on the day following the Publicans' Purse of twenty-five sovereigns fell to his lot. At Homebush May Meeting, for the All-aged Stakes of fifty sovereigns, he distanced the great southern crack, Petrel, although Munchausen rumours had been circulated as to the wonderful capabilities of the Port Philip favourite. The Australian Plate of fifty sovereigns next came to his owner's share, in defiance of a large field, including Lottery, Green Mantle, Tamarind, Emerald, and Blue Bonnet. After running second to Emerald for the Town Plate at Maitland, he surprised the widos by taking a Purse of sovereigns from the Irishman and Meteor, and further receiving an accession in the shape of the Hunter River Stakes. At Five Dock he walked off with the Ladies' Purse of forty sovs. in March, and shut up shop for the season.

"The March of '47 brought the Bathurst Town Plate of 50 sovs. and the Publicans' Purse of 30 sovs. April, too, added the Members' Purse of 60 sovs, and the Richmond Purse of 40 sovs., both at Windsor. In May, the Australian Plate of 60 sovs., at Homebush, brought more grist to the mill, and

June yielded fruits of increase—viz., the Parramatta Town Plate of 60 sovs., snatched from Whalebone and Blue Bonnet; and on the 2nd of July he won the Cumberland Turf Purse of 60 sovs. Dull December had also its attractions, as he was allowed to walk over for the Ladies' Purse (25) at Petersham, and on the same day he took the Hunters' Stakes of 20 sovs. In 1848 his prowess was rewarded with a score at the Drapers' Club; £50 (Publicans' Purse) at Bathurst, in March; a like amount at Windsor (Members' Purse); £50 (Town Plate) at Parramatta, in April; and £70 (Australian Plate) at Homebush, in May. Early in March, '49, we find him walking off with the Carcoar Town Plate of £50, and at the close of the same month a similar amount was awarded him at Bathurst. Then followed, in April, the Richmond Purse (£50) and the All-aged Stakes at Homebush; a sweep of 5 sovs. each, with 50 added in May; allowing one day to intervene, he bore off the Australian Plate of £70, with a sweep of 5 sovs. each, (a dead heat between the Old'un and the Plover, when the latter and other horses were drawn, and Jorrocks walked over for the second heat). The Cumberland Turf Meeting next came to the relief, in the shape of the Parramatta Town Plate (£50), with a sweep of 5 sovs. In the following month he walked over for the Wellington Plate (£20), and the

Ladies' Purse of £20, and closed the year at Mudgee, on the 15th of August, by carrying off a plate of 30 sovs. ; and thus he closed the year '49.

"1850—February 27th, ran second for the Town Plate of £50 at the Bathurst Meeting, being beaten by Little John—winning the first heat, running a dead heat the second, being beaten for the third, and then drawn.

"March 15th, at the Carcoar Meet, won the Publicans' Purse of 30 sovs., beating Little John and Zingaree.

"April 10th, at the Hawkesbury Turf Club Meet, won the Members' Purse of 50 sovs., ridden by Cutts, carrying 9st. 3lbs., beating Humming Bird and Pastile.

"May 29th, at the Homebush Meet, ran second to Sultan for the All-aged Stakes of £50, beating Little John and Plover.

"May 31st, same meet, won the Australian Plate of 60 sovs., with a sweep of 5 sovs. each, beating Plover, Little John, Lola Montes, and Sultan.

"27th June, at the Campbelltown Meet, ran second to Plover for the Campbelltown Plate of 50 sovs., beating Pasha, Sultan, Pastile, Sir Charles, and Jenny Lind.

"August 6th, at the Singleton Races, won the Patrick's Plains Purse of 40 sovs., beating Sultan, Death, and Cyrus.

"August 20th, at Maitland Meet, ran second to the Queen of Hearts for the Maitland Town Plate of 50 sovs., and on the 22nd, same meet, ran second to Sultan for the Town Plate, which wound up his career of 1850.

"1851—Feb. 26, at Bathurst Races, ran second to Little John for the Town Plate.

"Feb. 28, same meet, won the Publicans' Purse of £40, with a sweep of £3, beating Little John.

"April 25, at Hawkesbury Races, was beaten by Plover and Muleyson for the Richmond Purse.

"May 28, at Homebush Meeting, for the Australian Plate ran third, being beaten by Cossack and Muleyson.

"Jorrocks then left the Turf, and after several times being raffled, became the property of his now owner, Mr. A. Thompson, who once brought him to the post, on the 7th of Oct., '52, for the Metropolitan Handicap, at the Drapers' Homebush, when he was beaten. A most marked manifestation of interest was displayed as the "old horse" was led upon the course, and three deafening cheers made the welkin ring again as he passed before the stand. Poor old horse! the spirit of other days was unextinguished— is unextinguishable; but no new trophies may be gathered for his laureled brow. His fame is imperishable, and the records of past years will advance his undisputed claim to the first niche in the Gallery of Australia's Winning Horses."

But Racers as racers form no part of my subject; as to their adaptability to be sires of saddle-horses, I have already shown it to be against all principle; but principle aside we can judge of the fruit without any discussion about the tree.

Taken as a whole, the saddle-horses of Victoria, New South Wales, and Queensland, of which I can speak from personal observation, and to which, I believe, those of Adelaide offer no exception, are not difficult to describe. On the whole, they are in height decidedly below those of England. In their appearance in the streets of our principal cities, there is nothing that particularly strikes the English eye, except it be their unvarying inferiority of figure and utter want of quality. Year by year they have been subsiding and settling down to a dead level, and the bright exceptions which, twenty years ago, were frequent amongst them have now all but totally disappeared. This is more strictly the case in Melbourne than in Sydney. In Melbourne there may remain a few stylish pairs of carriage horses, but amongst the scores that are sold by auction every day in the week, perhaps even one tolerably good colt could hardly be picked out. Those waiting private sale are only a degree or two better, and, if enquiries are made for good-looking, well-bred sound hunters, or hacks, up to fourteen stone, in the places where such used to be met with, the enquirer will at once be told that such horses are now never to be met with. What is noticeable in the capitals is but the reflex of the whole colonies, and in corresponding degrees. Thus the horses of New South Wales are more presentable than those of Victoria.

A cessation of this decline, and a gradual but permanent rise to excellence, has been foretold, believed in, and hoped for by many. Twenty years ago I used to hear that Cornborough, Patchwork (who died on landing), and other horses whose names it is useless to mention, were to regenerate and bring up our breed to the old mark and higher. This delusive hope has always found its votaries—at present they are dreaming of Indian Warrior and Fisherman. They, it is confidently asserted, will inaugurate the millennium—and when they have passed away, and effected no more than their predecessors, fresh phantoms will no doubt replace those that have fled, and beguile with a hope which will certainly never be realized. Though, were it realized, it would still, I maintain, be far less, both in degree and extent, than what is easily possible under a proper system of breeding. Passing from the figure to the qualities of the Australian horse, it will be found, that in comparison with the English horse, he is ill-broken, his temper cross, and his paces disagreeable; that, if compared to the horses of the Arabs, or those of Southern Italy, Greece, Asia Minor, Palestine, or Egypt, he will be found to be, besides, sluggish in temper, unsound in his legs, soft in his hoof, and wanting in stamina.

After all, however, there is one favourable feature still noticeable amongst our horses, and it is a most important one; it corroborates what we hear of them in the past, and gives us confidence in the future. I mean *their great capacity for work, as compared to their figure.* In this they still, and always will, rank high, for it rests on what is unchangeable, and beyond our control, as I shall endeavour to make clear in the sequel. In this quality they surpass many others of even a better figure than themselves, and I feel no hesitation in saying, that if an English horse in England, for instance, of a certain figure and breeding, was found on trial to perform say thirty miles a day for a fortnight, with stable feed, as the maximum of his capabilities, that an Australian horse of exactly the same figure and breeding, with no other feed than grass, would certainly perform forty-five miles a day for the same period. This superior stoutness of the Australian horse *according to his figure*, as compared with the English horse, is pretty generally acknowledged by those who have had experience of the horses of the two countries. Thus, a hundred miles, on an emergency, are frequently being done in Australia, by very miserable looking Rossinantes, in fifteen hours, without preparation and off grass. Eighty miles two days consecutively, and seventy miles three or four days following, are constantly being done in the routine of business, in like manner. I myself have done these things and more.

In a journey of 400 miles, which I have several times had to perform, I have started with two fat horses unused to work (so far, that though well seasoned, neither of them had probably been backed for a month previous), riding one and leading the other, with a small pack on his back, changing the saddle occasionally from one to the other. The 400 miles were always accomplished without trouble in eight days, and after three days' rest, the horses were quite fit to return at the same speed. My business, however, usually detained me from a week to a fortnight in town, when I returned as I came, always doing the last 50 miles by one o'clock, a.m., and turning out my horses somewhat weary, but with plenty of fat and flesh about them, ready to repeat the same journey after a week's rest. At night, they were hobbled out, usually on very scanty grass, and never tasted artificial feed of any sort. The object in taking two horses, which, in New South Wales, as at the Cape amongst the Boers, is frequently done, was because such paddocks as existed at the stations on the road were quite destitute of grass, which obliged me to camp out, and in camping out few horses, when fresh, will stop alone. Two in company usually stop well, whereas one, however tired, will often wander ten or fifteen miles during the night, even in hobbles, and perhaps not be found for a week. Even when camped out, the grass for the first 250 miles was excessively scarce, much more so than is ever known in Victoria, for instance.

These I give as instances of ordinary work on fair horses. One of those which I rode was in appearance little better than a cart-horse, the other an old mare, well bred, but, as it turned out on the last occasion, four or five months gone with foal. This mare, Nettle, one of Mr. Wyndham's, of the M'Intyre's breed, would have done the whole journey alone at the rate of 70 miles a day, in her palmy days, I feel no doubt. I rode $13\frac{1}{2}$ stone on these occasions, and could have done 100 miles the last day in fifteen hours, had I desired it.

In overlanding with cattle, it is usual to allow the men employed to drive the cattle, two horses each. The work is done at the slowest possible walk, the merest crawl, but the horse is under saddle at least 12 hours on the stretch, and is ridden and spelled alternate days. This often lasts for six months incessantly, and if the feed is good, the horses will fatten; if bad, they become very poor.

Twelve years ago a pack of foxhounds used to be kept on the Murray, not far from Moama, with which I occasionally hunted. We all, except the whip, rode grass horses. The wild dog generally lived eight or ten miles before the hounds ran in to him. There was seldom more than one check, and I never saw a horse left behind, or appear very much distressed. What the pace would be I cannot say; the country was a dead level, and there were no jumps; in some parts the ground being very hard, at others deep and loose sand; the scent was often all that could be desired, and I believe there can be no doubt that the wild dog is more speedy than the fox. He makes no turns in his course, if he can avoid it; but goes right ahead.

For a couple of seasons, if the reader will pardon a digression, the author kept eight or ten coarse greyhounds or kangaroo dogs. Probably none of them were quite pure, though they were generally pretty well bred. In point of speed, in a straight-on-end run of two or three miles, such as the kangaroo and emu generally give, I do not think they would have been easily passed by any greyhounds. These dogs, kennelled like foxhounds, soon *learned to hunt in a pack, and by practice, with the nose quite as well as by sight,* and were not slow in acquiring that quick perception of each others' movements, when hunting a cold scent, which is so noticeable in the foxhound. The finer or more greyhound-like the individual, the more acutely developed, I noticed, when cultivated, became the sense of smell. Their noses arrived at a degree of perfection which quite surprised me, having always understood that the greyhound is almost devoid of that sense. They stood about four inches higher than the generality of English greyhounds, the largest pups having always been selected from each litter, for several generations, and their food, which was always meat, given without stint.

These kangaroo dogs killed 60 wild dogs the last year I had them, besides many kangaroo and emu. Their pace on the scent was considerably better than that of foxhounds, the only difference perceptible in their mode of hunting, under any circumstances, being that they were of course silent. I often followed them seven miles, in all sorts of weather, hot, cold, and dry, over thick sheep tracks, as hard as I could rattle, before sighting the quarry, and sometimes as much as ten miles, blowing my horn from the start; my grass horse was never violently distressed, always less so than the dogs, if the weather were hot. After killing I took a few minutes' rest, and sometimes had a second run on my way back. On reaching home, the saddle and bridle were at once removed, and the horse allowed to return to grass and water. Were greyhounds in England to hunt by scent as well as sight, they would soon exterminate the hare, for there is no doubt that if the use of the nose was cultivated in the pure descendants of the finest greyhounds in England for a few generations, they would rival the foxhound in his line. Such is found to be the case with the slougui of the Sahara.

I saw these dogs of mine run the scent of an emu in blazing hot weather, twenty-four hours after he had passed over dusty ground, entirely devoid of grass: when they had become adepts in their *trade*, no beast in the bush, with half an hour's start, either could or did ever escape them. On the plains, they used to amuse themselves with hunting quail, which were exceedingly numerous, as I allowed them to hunt anything but sheep. They seemed to scent them quite as well as a pointer, ran their trails, and sometimes caught them in their mouths as they rose. These kangaroo dogs were not remarkable, that I am aware of, over others, but being much hunted, and having much blood, they became very resolute, and would spare no pains with a dull scent. They were well known in the neighbourhood where Echuca now is: *mais revenons à nos moutons*.

Having said this much of the Australian horse

THE AUSTRALIAN SADDLE-HORSE. 173

in general, I must now hasten to direct the attention of my reader to the distinctive features of our horses, bred in different localities. As the horse is found here, inhabiting districts extending through many degrees of latitude, with much diversity of climate and pasture, it is natural to suppose that some results have already arisen from these causes, and become apparent, and such, in fact, is the case: thus, *horses bred on the sea coast, it is very worthy of remark and important to recollect, perform less, for their make or figure, than similar horses bred at a distance from the sea.* They are doubly indifferent where the country so situated is richly and heavily grassed: they then become tall, large, and fleshy, but are soft, and can support neither hunger, thirst, nor long journeys. In riding them to fox-hounds they require stable feed, and even then do not always see the finish. If taken into the interior country, which is infinitely hotter than the sea-coast country, they are almost incapable of work for two or three years, when they improve very much, but still remain inferior to those bred on the spot. Neither do they endure cold better than the horses bred in the hot country. Of these facts I have had plenty of experience in Victoria, New South Wales, and Queensland.

Of a distinct class, and a degree better than these,

are the horses bred in the mountains, at the heads of the rivers, and upon or near the dividing range, as likewise on the table-land of New England. These districts are known as "green grassed" or "sour" countries. Neither sheep nor cattle fatten well in them: hence they are used as breeding, and not as fattening countries. The horses bred in such situations have usually a greater show of muscle than those bred on the plains. They are surefooted from habit, and accustomed to hills, but soft, great drinkers, and shy of the hot sun. They cannot perform long journeys, or support continuous work, like those of the interior. With these must be classed those of my native land, Tasmania, as well as the majority of the New Zealand horses.

The worst horse in Australia is the Queensland sea-coast horse, and the nearer the tropics, the damper the locality, the more indifferent he becomes.

The best of our horses in proportion to his figure, the most abstemious, stout, and sound, and the most neglected in his breeding, drinks of the waters that flow into Lake Alexandrina: that is, the horses bred on the rivers Darling, Lachlan, Bogan, Murrumbidgee, and other tributaries of the Murray, after they leave the mountains, and before the Murray approaches the sea. These things I say from personal experience, except in the case of the Darling, on which river I have not been, and which I only speak of from report. I feel no doubt but that this will one day be well known, and that the time will come when the arid plains and burning hill sides of Central Australia, will produce a horse that will be famous all over the world, and perhaps surpass even the Arab of the desert. But on this head I shall touch more fully in the chapter on Climate.

Without, however, receding long distances from the coast, there are districts in which admirable horses will probably one day be bred. Victoria, for instance, from its salt-bush plains, lying between the Campaspe, Murray, and Richardson, will furnish itself with excellent horses, and New South Wales from similar country adjoining. From personal experience of these countries, I have no hesitation in saying that they are as certain to produce long-winded, stout, generous horses, whenever proper means shall be used to that end, as that such a result is in an equal degree impossible in the coast districts. Another and a bad characteristic of our horses is, the frequency of vicious tempers which is noticeable amongst them. On this subject, Captain Shakspeare, in his "Wild Sports of India," remarks: "They (the Cape horses) appear to be severally good tempered, much more so than the Australian and New South Wales horses, which used to be quite unbroken and almost unmanageable when first sent over. They have very much distinguished themselves as racers and carriage-horses, but otherwise I consider them to be the most difficult horses to break that can be found."

This is no matter of wonder, and proceeds, I am inclined to think, from improper breaking, which, continued from year to year, and from generation to generation, amounts to nothing more than a cultivation of vice, on just the same principle that docility or any other quality is found and admitted to be both susceptible of culture, and transmissible from parent to offspring. "For," says Stonehenge, "acquired qualities are transmitted, whether they belong to the sire or dam, and also both bodily and mental."

This defective breaking in Australia, which aggravates and hardens the natural vice of our horses, arises in part from the high price of labour, and the ignorance of many of our breakers : for all who can ride here consider themselves competent to break, and do break when opportunity offers. In other and very frequent cases, especially in the old bush districts of New South Wales, this faulty breaking comes from another cause : the pleasure experienced by the rider in mastering a savage horse in his angry mood. His object is not to quiet, pacify, and break his horse, but to make him do his worst, and sit him. Perhaps there are no people under the sun who can sit a vicious horse like the Australian-born Englishmen of the bush district of New South Wales. After several generations, climate, food, and habit have, as usual, left their impress on the figure. The bush-born *natives*, as they are called, of New South Wales, are, as a rule, tall, slight, and neat-figured, and very rarely become corpulent. Their arms and legs are wiry, and inclined to be long; shoulders and trunk light. They are muscular, abstemious, and active, and decidedly a sober race. Few of them use tobacco. They walk well when practiced, and endure hunger, heat, and thirst better than the emigrant, who surpasses them in his turn in toils, where strength is a requisite, and weight no detriment. They are taciturn, shy, ignorant, and incurious—undemonstrative, but orderly, hospitable, courageous, cool, and sensible. These men ride like Centaurs. Ask one of them to ride the horse that has just thrown you: he examines the girths, crupper, and bridle. His face shows no emotion of any sort. If the tackle is right, he lifts his hat, lets the string fall under his chin as he replaces it, carelessly gathers up the reins, and mounts. When once in the saddle he sticks there, and enjoys the row in his own quiet way, and dismounts as unmoved as he got on, perhaps with some quiet remarks upon saddles or horses.

It is the custom in Australia, in the bush, to ride

vicious horses, as all others, in English saddles, and with snaffle bridles only. A curb of any sort is seldom seen amongst the natives. In breaking or rough-riding, a crupper is always used, as without it the saddle cannot be made to remain on the back of the buck-jumping horse. In the case of a low-withered horse, the saddle is forced on the neck, if the wither be high, the girths are usually burst if no crupper be used. Neither is the crupper strap, in all districts, attached to the crupper-iron or staple, which, in the event of a severe struggle, is certain to be pulled out or broken; but, being passed between the saddle and saddle-cloth until it obtrudes before the pummel, it is then passed two or three times round a stick, about twenty inches long, and as thick as the wrist, called the "kid," and there buckled, which kid is strapped to two iron staples, with which the saddle is provided, behind the pummel. This kid comes across the horseman about six inches above the knees, and adds very much to the security of his seat. All those, however, who are considered good horsemen will stick to the saddle without this or any other aid, as long as the saddle sticks to the horse.

One of the results of this excellent horsemanship, added to the high price of labour, which would make the English breaking more costly than the colt himself, is this, that many encourage their horses to buck for the pleasure of sitting them. From these several causes, the disposition has been cultivated amongst them, and has, of course, become constitutional and hereditary. As a proof that such is the case, if any were needed, it is noticeable that the buck-jump is the favourite gambol of the young foal, who goes through the motion admirably when he is six weeks old, and thus practises from his infancy that which he exhibits in such perfection when he comes to be backed. Every species of vice, except buck-jumping, is quite disregarded in the bush, and a horse that does not buck is consequently considered to be quiet; kicking, plunging, rearing, bolting, &c., clearly showing that the horse, though excited, is only playful, or he would at once have had recourse to bucking—as the only way any sensible nag could expect to eject his rider.

It was at one time supposed that the South American Gaucho, so celebrated for his horsemanship, would not only be able to sit our horses, but also to render them quiet, in some wonderful manner, at the end of a few lessons, and fit for the town markets. With this object, some years back, a few picked Gauchos were imported here, but not only were they unable to break our then well-bred, vigorous, sixteen-hand horses, as they do those of their own country, but when our snaffles were put into their hands, instead of the jaw-breaking curbs which they had been accustomed to, the falls which they got were frequent and severe, and after some months the trial was abandoned as a failure.

It may not be uninteresting to mention, that the Australian aborigines, when practiced, become excellent horsemen in some respects. Almost any black, between the age of fifteen and thirty, will, at the end of a month's practice on a quiet horse, sit any unbroken colt, and most of them, at the end of another month's practice with buck-jumpers, will ride anything that can be produced. Their other strong point is as scrub-riders, in which, on the whole, perhaps they are unequalled. The universality with which they excel, when allowed to try, is very remarkable. Where they are employed as stockmen and rough-riders, they pride themselves very much on their performance. The wonderful part of the business is how they get on with their horses. They are quiet, but not familiar—seldom irritating, and never caressing them. A horse broken by a blackfellow has always a wretched mouth, will never lead, and is usually very shy of objects with which he has not been rendered familiar. His education is very limited. They all ride with short stirrups; rough-riding and scrub-riding are the only points in which they excel when mounted. As race-riders, in which, at country meetings, they are a good deal employed as light weights, they are very indifferent. They enjoy the sport, ride with a loose rein, and are very apt to make play with the spur from first to finish if allowed.

Horses broken by them seldom walk well, and rarely trot at all. They look well on their horse; and in the points spoken of, seem all to excel. With us the reverse is the case—for many white men can never learn to ride, however much practice they may have. The fact is their figure and nerve suits the occupation, but intellect, and above all care, are wanting.

In all of the Australian colonies there are many proprietors of large "mobs" (I cannot call them studs) of horses which are allowed to go at large, and are all but totally neglected. The stallions run loose with them, as in the wild state: the utmost labor expended on them is to run them into the stockyard now and then when some of their number are needed for work. Of studs (except a few small ones), round the several capitals, in each of which a score or two of racing stock are bred, there are not many; and several that have been famous have either been broken up or allowed to deteriorate from neglect.

The only stud of saddle-horses with which I am personally acquainted is that of Mr. Wyndham, of the M'Intyre River, New South Wales, which bears the reputation of being the second, if not the best in these colonies. From this stud have come Lunelle and Laurestina, winners at Homebush, and other racers of note, which have been, however, as is proper, only accidental in a race of saddle-horses. Their blood being that of Snake, Scratch, Plover, &c., thoroughbred horses of character and power, their figures as a consequence are good, as times go, with, however, a short shoulder as a family failing. Having the legs, pluck, and other requisites of good saddle-horses, as compared with the Arab horse, in a moderate degree, they are, however, much wanting in style. Their heads are somewhat coarse, the eye often small and pigish. Having some years ago been the purchaser of a hundred mares from this stud, most of which, when broken, I sold at Nelson, New Zealand, I had a fair opportunity of judging of their merits as saddle-horses. Their legs, as I have said, compared with other Australian horses, were good; they were high-couraged mares, with pluck to face anything. They leaped readily and well, were free, pleasant and easy goers, and looked better when mounted than when loose; they were sure-footed and long-winded, excelling in the canter and galop, ordinary at a walk, and bad trotters. They carried fourteen stone well. Their tempers were all hot, many of them cross, and not a few unmitigated buck-jumpers, though considerable trouble was taken in their breaking. Taking them altogether they were decidedly, and by far, the best 100 saddle-horses I ever saw together in the colonies. As regards soundness, they were much above the average of our horses here, but as regards stoutness, *in comparison with their figure*, I cannot say so much. Though decidedly very stout horses (much above what any cavalry regiment in England could turn out), they were quite as certainly not what they should have been. This is not difficult to account for. Very old worn-out sires had been much used, and mares retained in the stud ten years after they should have been culled. Besides this, and worse, they were bred in a cool, timbered, half-sour granitic country. Horses from such soil and climate are not the real grit, they never perform what horses of their figure are capable of in other climates. I do not believe that one really first-class horse was ever the product of a granite soil, or of a cool climate, or ever will be. Be this as it may, I speak of the mares as I found them.

In reading the *Turf*, by Nimrod, I could not be but struck by the following passage, which at once brought to my mind one of the same illustrious family here spoken of, and attending here, almost singly, to the *legitimate end of racing*. Should this meet his eye, I trust he will excuse the liberty I have taken with his name, and receive in a friendly spirit my remarks on his stud.

"But we cannot say this of the noble earls, amongst whom are some of the best judges of racing of past or present days. We will begin with the Earl of Egremont; and not only by the rule of *seniores priores*, but looking on him as one of the main contributors to the *legitimate* end of racing— *the improvement of the breed of horses*—his lordship having always paid regard to what is termed stout, or *honest* blood. Lord Egremont bred Gohanna, by Mercury, by Eclipse, and purchased Whalebone from the Duke of Grafton (the old Prunella sort), whose stock have been invaluable to the turf, and will continue to be so for many years to come, although objections are made to their size— made amends for, in great measure, by their symmetry. His lordship has likewise turned the amusement—and such has been his main object in the pursuit of it—to an excellent account, in the liberal act of affording to his tenantry and neighbours the free benefit of several of his stud-horses. Among these have been two very fine animals, Octavius and Wanderer, the latter not inaptly named, as for many years of his life be was never known to lie down, but was generally in action in his box. He was a noble specimen of the horse, and one of the best bred ones in the world, for all the purposes for which horses of speed and strength are wanted, being by Gohanna, out of a sister to Colibri, by Woodpecker, esteemed our stoutest blood. The earl is likewise the breeder of honest Chateau Margaux and Camel, ornaments to the British turf, and sons of good little Whalebone. Lord Egremont won the Derby three times in four years; twice with sons of Gohanna, subsequently with Lapdog, by Whalebone. He has also been three times the winner of the Oaks, with fillies from his own stud. But all this success is not to be placed to his lordship's own account; he received great assistance in all his racing speculations from his late brother, the Honourable Charles Wyndham, since whose decease the stable has not been so successful."

In Australia, as in America, we have considerable herds of wild horses, the offspring of such as have at various times escaped from stations and remained for years undiscovered or unyarded. It is, of course, quite impossible to say to what their numbers may amount, but it is probable that they do not fall short of 25,000. In the neighbourhood of stations where they exist, they have become a serious evil, frequently enticing away the domesticated horses, which are rarely recovered. As the rule, wild horses that are captured are found to be small, light-boned, weedy, and useless, the effects of early breeding and hap-hazard sires. In temper, they lose some of the active vice of the domesticated horse, manifesting in its stead a species of dogged obstinacy and stupidity. In the terrible chase which precedes their capture, they generally become heart-broken, whence it frequently happens that a horse which will gallop fifty miles before he will allow himself to be yarded, can, when broken, be neither persuaded nor coerced to carry his rider five miles, though if again let loose his ability to gallop, and his determination to be free, will be found to be undiminished. Resignation, generosity, and emulation, seem frequently quite wanting in them. They tremble when bridled, and resist and resent all caresses, and every step they take when mounted appears to be *contre-cœur*.

In figure they are small, ill-shaped, under-boned, large-headed. As they are very shy, they seldom approach a river, except when compelled to do so, frequenting as long as possible sequestered springs and waterholes. When these are exhausted in dry weather, and they are compelled to drink at the rivers whose banks are inhabited, they usually do so at night, making their approaches in scrubby and rugged localities, coming and returning at a trot, which the least alarm at once converts into a headlong gallop. So great is their dislike to approach localities frequented by man, that their visits to the rivers are habitually delayed, till their thirst can no longer be borne. This habit of long abstinence from water, has given them a tucked-up appearance, which even when domesticated and allowed to get fat, they never lose entirely. Amongst many other places, there are large herds of them on the lower Lachlan and Murrumbidgee rivers. In these districts, on the flat, treeless, horizon-bounded plains, one would suppose it an easy matter to stockyard them, but it is not so.

In running them in, many foals and weakly ones are left behind, knocked up, and it generally happens, especially in hot weather, that of those yarded several die in a few hours from the effects of the gallop.

"Running" wild horses, as the process of galloping them into the stockyards is termed, is both exciting and dangerous in the extreme, as well as expensive; and rarely remunerates the undertaker, yet, like other lotteries, some people find a charm in it. The country which the wild horses select as their haunt, is always the most broken in the neighbourhood, if there be a choice; and if they should be found in more manageable country, they make off at once, on being disturbed, at headlong speed for their chosen retreat. That those who pursue them should frequently come to grief, is not to be wondered at, when the pace at which they must go, and the length of time they are often obliged to keep it up, are considered; hence accidents are very common.

The systems followed in running wild horses are various; the one adopted depends on the nature of the country in question. In some situations, two or more lines of fences, extending perhaps three or four miles from the stockyard, and termed "guides" are erected. These lines, commencing at points two or three miles apart, gradually converging, meet at the gate of the yard. The fugitives once between the guides, pursued closely by the riders, stock-whip in hand, have no choice but to enter the stockyard, whence there is no escape when the gate is once closed.

Where there are no guides, relays of horsemen are posted here and there in positions where the herd is expected to pass, ready to take up the chase with fresh horses. But, under the best arrangements, it not unfrequently happens, that after a hard gallop, two or three valuable saddle-horses are broken down, and the whole wild herd effects its escape.

Losses more serious than this, even, sometimes occur; as in the following instance. A person wishing to get a lot wild horses, having arranged with the settler on whose run they were, for the use of his stockyard, and permission to run them (for they are usually claimed by some one, however illegally, as I believe ownership can in no instance be established), took his measures as follows. Having hired half a dozen or more riders, he purchased some superior saddle-horses, in the proportion of two or three to each rider; and besides these a mob of about 60 or 70 half-quiet, cheap, weedy horses, commonly called *crawlers*, to be used as a decoy mob. The country in which he was going to run the wild horses consisted of flat plains of unlimited extent, and almost without trees. All being ready, he started with his mounted men, driving his crawlers before him. Having got eight or ten miles from the stockyard he had left, and near the feeding grounds of the wild herds, he put his crawlers in charge of two of his men, who, remaining near them, allowed them to feed, but kept them from leaving the spot selected; he himself proceeding with the rest of his party in search of a wild mob. Such he was not long in finding, and luckily got them headed at starting in the desired direction. Down they came racing, and, sighting another mob before them, made straight for it, as their custom is. This mob was the crawlers, and received the strangers without becoming too much excited. The wild horses, finding that their new comrades did not join in the flight, refused to fraternise with them, and galloped through them as they were spread out on the plain. Then came the time for action for the two men left in charge. They, galloping a-head shouting and cracking their whips, and whirling their long lashes over their heads, stopped or "blocked" as it is termed the wild horses, turning them back on the quiet ones. By this time had arrived the rest of the party, who, circling their horses round and round the frightened mob, now neighing, biting, and galloping round amidst clouds of dust, backing up their comrades, succeeded at length in steadying and calming the disturbed body. Finding escape impracticable, their flight blocked in all directions, and a break impossible, the wild gradually began to fraternise with the tame horses, and after much trouble and time had been allowed for their excitement to subside, the whole body was put in motion, some horsemen riding a-head to restrain the leaders as much as possible. Reeking with foam, and covered with dust, the stockyard was fortunately gained, and, the most difficult point, entered without accident.

On drafting the wild horses from the crawlers, it was found that sixty had been bagged. These sixty were kept in the yard two days without food or water, in order that they might be weakened and reduced to a more manageable state, for had they been let out of the yard before they were much exhausted, they would have been so excited by the strange things they had seen, that they would have broken away, mad with fear, in spite of opposition, and regained their freedom.

A couple of days' hunger and thirst having reduced them to reason, they were again joined to the crawlers, watered, and taken out with them on to the plains to assist in the capture of a fresh mob in the manner before described. A fresh lot of about fifty wild ones was not long in being got, joined to the mixed lot, steadied without much trouble, they readily taking up with their former friends in freedom; which was the more easily effected, because the more numerous the decoy mob, the sooner the new arrivals lose their impetus, and get separated and confounded with them.

In the meantime, a large herd of about 150 wild horses had been seen, and the conductor of the hunt, too eager for gain, instead of being satisfied with what he had captured and taking them to the yard to be subdued by thirst and hunger, as he had done in the first instance, determined to take this other lot also with him. Steadying his former captives and leaving what men he could spare in charge of them, he moved off with the rest of his party, and was not long in falling in with and heading down the other 150. Away they came in the direction of his depôt party, each as hard as he could rattle, in a long string; the stallions and strong horses leading; the mares, the old, the lame, and the foals, stringing out, but following the lead at their best pace. On their approach, those that had only just been joined to the crawlers, now well in wind, began to cock their ears, whinny, and become restless. Their anxiety communicated itself to the lot first captured, and, as these feelings of alarm are very contagious amongst gregarious animals, even the wretched crawlers began to get uneasy. In the mean time, the tramp of feet grew louder, and the 150 drew near, racing over the plain, mad with excitement, their tails streaming in the air and clouds of dust flying from their hoofs, with a thunder of feet that might be heard miles off. Such was their impetuosity, that the quiescent body of half, subdued material was unable to stop them, and the whole affair became a rout, and every horse there, to the poorest crawler, joining in the rush was carried away by the excitement of the moment. Round went the two lots at meeting, in whirlwinds of dust; amidst its mantling clouds, the tossing of manes and streaming of tails, a thousand incidents flash indistinctly for a moment on the eye of the beholder; the tall grey that has run headlong against another in mid career, rearing high above the rest, for an instant his full height, tumbles backward with his neck broken; the vicious old mare, that hates a throng, with head bowed to the ground, kicking all that come within reach ; the foal, overthrown, rolling amongst the feet of the multitude; the neigh—the savage yell—the scream of terror—the thunder of hoofs—the rival stallions meeting in the eddy, with outstretched neck, snake-like head, and ears laid back, true to their nature, springing with fury on each other, to satisfy in a moment of mortal terror the jealousy and hate which never brook postponement of battle; all had been and disappeared in an instant, for nothing could restrain them. Bursting from their captors, blind and deaf to danger, headed by some resolute stallions refusing to be turned, the whole lot broke from control and swept over the plains. The yells and cannon-like reports of the whips of the pursuers only added fuel to the fire. The whole lot escaped, and were soon out of sight behind the swells of the plain, the cloud of dense dust floating in the dry air marking the direction of their flight, the person at whose expense the hunt had been got up having to return home with his party, losing the whole of his horses except those on which he and his men were mounted, with the pleasant reflection that the domesticated horse, once free and joined with wild ones, is not a whit easier to recapture than the wildest denizen of the plain.

In many parts of the country these large herds of wild horses have become a serious inconvenience to the settler, and is one which is always on the increase. As a body they are much inferior in figure, symmetry, and quality to their domesticated brother, though a pretty good-looking animal has now and then been got from amongst them. Their destiny eventually, when the country becomes fenced in so that they can be captured wholesale, will no doubt be the boiling-down establishment and the tallow-pot. For any other purpose they may be said to be useless. I believe some have already been boiled down, but with what results I am not able to say.

Before closing this chapter, it may not be amiss to say a few words concerning the export of horses for cavalry purposes from Australia to India, during the late mutiny. Whatever *prestige* our horses may have heretofore enjoyed in that country will, it is more than probable, have been entirely dispelled by this last export. It was a necessary consequence, and to be expected, from the degeneracy of our horses, that the late cavalry remounts should in quality have fallen very far short of what we once supplied, but there was still no reason why the dearest and least fitted for the climate of India should have been selected. No doubt, as compared with cavalry management generally, it was congruous enough, and so fails to be a matter of surprise. Sending to Australia for horses is like sending to Europe. Australia embraces many climates, many soils, and horses of various qualifications. Many, or I believe all, of those selected, were from the neighbourhoods of Sydney or Melbourne. They were horses of less stamina than are found farther from the coast, and, from the nature of the climate where bred particularly ill-suited to India. It is, further, not unlikely that many of them are notoriously vicious, and therefore useless for cavalry in action, whatever they may be on parade. No doubt that a large proportion of them will have been found to be unsound. In my judgment, it must have been a mistake to purchase broken horses at all, excepting such as might be required for immediate action. Then, in what sort of vessels were they shipped, and what was the quality of forage found for them? I myself know an instance, where almost a whole cargo was lost, from the nature of the fittings, which all gave way on the first gale of wind, and the bad quality of the hay; and both of these circumstances were pointed out to the officer superintending the shipment before the vessel sailed. These facts were afterwards proved at Calcutta. Would any private speculator have been so careless and improvident? It would be interesting to know what our horses cost the Government a-head on landing in India.

If the Indian Government ever again require Australian horses, and they send an officer into each of the interior districts, where horses are numerous, and stout in constitution, whose business it should be to inspect at the various stations such horses as are offered for sale, branding those that suit, and allowing the owner to ship them to India at his own risk and cost, payment to be made for them at a fixed rate on delivery in Calcutta, I feel sure that better horses would be obtained, and in better condition, at half the cost of the late shipments. For it it would seem quite clear, that individuals, having hundreds of horses, cannot drive them to Sydney and Melbourne for inspection, incurring the risk of but few or perhaps none being purchased. To drive 200 or 300 horses, say 300 miles, is both troublesome and expensive; for an officer to ride that distance to look at them should be a matter of small moment. As things were managed, every detail of the transaction probably cost the Government twice the sum that it would have done persons accustomed to business and living on the spot. The horses which thus reached India were neither a fair sample of our horses, nor of the cost at which we can furnish them to that country.

PART II.

"Speak your latent conviction, and it shall be the universal sense
.... Else, to-morrow a stranger will say with masterly good sense precisely what we have thought and felt all the time, and we shall be forced to take with shame our own opinion from another."

Emerson.

ON BLOOD.

"Blood is here synonymous with breed."
Stonehenge.

I shall now proceed, gentle reader, to speak of the principles of which the foregoing pages are meant to be the illustration. My reader is perhaps a breeder of horses : has he ever cast aside received opinions for the moment and quietly reflected on the subject for himself? Has he ever weighed the fact that the racer is the offspring of a racing sire and a racing mare ; that no horse bred otherwise is a racer ; that the racer can be bred in no other way ; and that a horse bred in any other way cannot race ? Does he know that experience has shown and writers told us, that departure in practice, however slight, from this theory has always ended in a marked failure ? Has my reader noticed that good draught-horses come only of stock long famous in the collar ? Is he aware that Shetland ponies never come of Clydesdale mares, and that the thoroughbred mare never drops a foal that grows to be a brewer's horse? Probably such facts may have glanced through his mind, and he may have noticed in a general way that each seems to produce after its kind, and as it were transmit itself to its offspring. It may further have struck him that this tendency is not an exception to, but rather a marked propensity and fundamental law of, nature; that it is evident in the human race—is exemplified in every variety of the animal creation, extending even to the lowest reptile. That it prevails as well in the vegetable as in the animal economy of nature, and is so constantly being exemplified before our eyes, as to lead us to anticipate it, and calculate on its recurrence as a certainty. So we have come to reckon on the growth of oak trees from acorns; and the sight of olive trees springing from apple pips, or any other such deviation from the worn path of nature, would certainly fill us with wonder. And not only have we seen these things, but some of man's actions have from time immemorial been grounded on what he has thus learned to expect. Hence he sows that which he would reap. If he wants wheat he sows wheat, and not barley, for he believes that wheat will not grow from barley, nor from any other seed or mixture of seeds, but from wheat only.

This truth, you will say, is threadbare—too trite

for repetition, assuring me that "like begets like"—that everybody knows "that what's bred in the bone will appear in the meat," &c., producing, as it were, fifty other rings in whose golden circlet is set this jewel *blood*. How inexcusable, then, must be the folly or ignorance of those who ignore this truth !

We admit, then, that wheat is grown from wheat. We are aware that the racer springs from the racer, the pony from the pony, the draught horse from the draught horse, and that each so bred has long proved immeasurably the fittest for his task. How then is it, I must persist in asking, that *our saddle-horse alone is not bred from the saddle-horse?* Why is our saddle-horse a bastard—systematically cross-bred, and that the very name of a *pure saddle-horse* is unknown, unheard of, and amounts in the very term to an incongruity in our language ? I should like to hear this question answered. On what principle have we acted ? Is any one bold enough to say that our action in this matter is referable to any principle of nature, or receives any support from practice ; and if so, what principle is it which favors this mode of action, or where are the results which justify this departure from principle ? Has nature in this instance departed from her rule, instituted some exceptional law—some "act to amend an act"—and refused to allow the saddle-horse, like others, to bequeath to his progeny his excellencies or his faults? Or have we been so ill-advised as to seek, forsooth, to improve upon nature, limiting and modifying her enactments? Or is the practice part of the residuum of kindred errors, which have one by one been passing away? Is it not a gross folly and inconsistency in our system, and can the length of time it has been tolerated be reasonably held up as an argument for its continuance? Would it not be wise and consistent to return to the natural laws on this matter, and trust to judicious selection and the coupling of horses themselves excelling in the saddle for a progressive amelioration and ultimate perfection of this kind? Would not this course be more consonant to sound reason than the one now in vogue, the expectation of powers and fitness in the offspring which are non-existent in the parents?

But are saddle-horses, bred purely from saddle-horses, really superior to those obtained from a cross of any description? Do facts bear out what I advance? I'll suppose my reader to say that "they are not." My question to him will then be—Do you speak of pure saddle-horses from experience? Have you ever tried those so bred—the Arab, the Barb, or the Persian Horses? If he says "Yes," which I deem very unlikely, I will beg him to weigh his single opinion against those of many others. But if he answers "No," I must then say, If your opinion be worth anything, as it does not rest on experience, it will probably be based on some evidence, and if so, what is it? produce it. I can only receive evidence of one sort, the opinion of competent judges who have fully tried the two classes of saddle-horses, the pure saddle-horse and the bastard saddle-horse at saddle-horse work; for I can no more receive racing results as a test of saddle-horse capabilities, than I can estimate the flight of the swallow by his speed of foot. If I prefer the pure saddle-horse of the East to the cross-bred of England, I do so from strong reasons, which I have adduced. First, as the result of well-known *theory* on the subject, then on personal experience and trial, and still more on the concurrent opinion, as far as I have been able to learn, of all those, without exception, who have tried the two sorts. Results, then, most fully tally with the theory we all admit; and are against our practice. Were results other than they are, what would become of our theory? I have accepted our theory that "like begets like," and my facts agree with it. The English saddle-horse, through the thoroughbred, is in part an Arab. He surpasses all such as have no Arab blood, and the pure Arab surpasses him, simply because the whole is greater than a part.

Whatever, then, the reader may think of breeding saddle-horses from other than the saddle-horse,

I cannot say. To my mind it is a folly, and I speak coolly when I say that I do not believe that one solitary first-class saddle-horse was ever bred in this way. But, strange to say, though we neglect purity of blood in at least one of our breeds, its necessity in all is most distinctly and elaborately laid down by most of our authorities on the subject. I have already extracted, at page 34, what Stonehenge says of blood, to which the reader will perhaps turn.

By reference to the passage indicated, the reader will see what Stonehenge says on the subject of *blood* when he addresses himself formally to the subject. Here, though perhaps not as precisely as might be desired, we find laid down for us the following most important facts.

1st.—That purity of blood, or breed, call it what you will, is only comparative, and means blood that comes from one source only.

2nd.—That a horse may be a pure racer, or a pure draft-horse, or a pure Naraganset, pure in fact of any class, or as it were for any purpose; and that this purity of blood and particular adaptability for any particular purpose is the result of constant uninterrupted selection of sires and dams suited for and excelling in the performance sought.

3rd.—That this quality of *blood, which is the result of selection*, is of all characteristics in a horse the most indispensable, the *sine quâ non;* which amounts to the full recognition of the fact, that any given qualities can exist in perfection in such individuals only whose progenitors and ancestors on both sides have, without exception, been long eminent for such certain qualities.

Now, if such is the case, any horse wanting in pure blood must be imperfect, he lacks the *sine quâ non*, and, as a consequence, saddle-horses bred on the English system of cross-breeding must be, and are, imperfect. Applying the principles laid down by Stonehenge, I should describe thus a pure saddle-horse : A horse may be said to be practically a pure saddle-horse, when it can be proved that his ancestors for fifty generations, or thereabouts, have been selected for individual excellence as saddle-horses, and from such families only as have been known to have excelled in the same particular. In two words, to produce a race of pure saddle-horses, breed for fifty generations with that design, strictly carrying out the principle *that like produces like, or the likeness of some ancestor.* This axiom embraces the whole science of blood ; it is consonant to reason and experience, susceptible of proof, and though disregarded by us in practice, admitted in the abstract, I believe, on all hands.

Stonehenge, as we have seen, always supports the

propriety of rearing cross-bred saddle-horses, though none other, as that measure is the only rational excuse for the large sum annually squandered on his darling the racer, and yet he tells us, "The fact really is, as proved by thousands of examples, that by putting *A* and *B* together, the produce is not necessarily made up of half of each." Again, "Experience tells us, that it is useless to expect to develope a new property or quality in the next generation by putting a female entirely deprived of it to a male which possesses it even in a marked degree." To be consistent in the matter, then, we must give up either our theory or our practice.

Though these things are clear, it may tend in practice to divest our minds of the very erroneous idea, that we can, in matching horses of incongruous powers and diverse fitnesses, fuse, as it were, and conglomerate the qualities of both, producing a useful combination in the offspring, if it be borne in mind, that frequently neither of the parents selected possess in reality any of the qualities sought. Thus, the breeder mates the racer, as recommended by Nimrod, with the draught mare, and anticipates from the union a hack possessing in reality the qualities of neither sire nor dam. For we must notice that the speed of the racer is of quite another sort from that which is required in a first-class saddle-horse, and that it is a poverty of our language which makes us express the two attributes by the same word. Speed in the racer is necessarily excessive in amount, and short-lived in duration; in the other, it is moderate in amount, but prolonged and lasting. Economize the pace of the racer, and you are still far from enabling him to work with the hack. The fuel is not saved in the same ratio that the steam is spared. Speed is cheap with the racer, endurance dear. So well is this principle understood, that Stonehenge tells us, that "It has been ascertained by experience that a horse (speaking of the racer) loses his pace for moderately short distances, if he is strained to the utmost for three or four miles."

Then, as regards the strength of the mare; she is strong enough to draw a ton and a half for twenty miles at the rate of two miles an hour, and yet not strong enough to carry herself along without encumbrance at the very moderate pace of six miles an hour, for three or four hours, or to gallop a mile or two, or leap a fence, or to exhibit many other proofs of strength which are found in the saddle-horse. Hers is somehow a wrong sort of strength for the saddle.

But in his estimate of blood, I find Stonehenge wanting in some particulars which are important; for, I think, he should have noticed, that as the health, stamina, and virtue of the horse, are in each successive individual intimately affected by the food, climate, and other circumstances in which he exists, so the horse of one country, pure from long continued selection, must still be inferior to those of only equal purity in another country more favoured in these adjuncts to perfection.

But in nothing are cross-bred horses so deficient as in endurance, in which particular they only illustrate a very general rule. For, whilst nature manifestly repudiates in-and-in breeding, she not less clearly sets her face against crosses. Her price for crosses is want of stamina. Hence it has been noticed in human races, as well as amongst animals of every sort, that half-castes, whilst usually excelling both parents in size, if not of too incongruous a parentage, are soft and unenduring, and that their offspring again diminish in size without proportionally increasing in hardihood. Those who would have the effects of pure blood without paying its price, loose the substance whilst grasping the shadow of perfection. What they get is cheap and bad. Endurance is pre-eminently the heritage of pure blood. "I am strongly inclined to believe," says Stonehenge, page 149, "that it is comparatively of little use to look about for sires who possess those qualities in which the dam is deficient." What I have advanced is so much in accordance, so thoroughly a consequence of the axiom of that acute author, that I can only attribute his stopping short where he does, failing to show the futility of crosses, to the influence of strong prejudices and partiality, of which he has been unable to divest his mind.

In seeking perfection in saddle-horse breeding, as in every other undertaking, the end as well as the means must be kept constantly in view, and in this instance it would be well to bear in mind the amount of perfection which it is possible to attain. An Australian does not know what a good horse should be able to perform, an Englishman far less. We look with distrust on tales of performances of Arab horses, just as a Bedouin would turn a deaf ear to our relations of railroad speed. He has never seen a railroad, the Englishman has never ridden a pure and good saddle-horse. Each is very ignorant in his own way. Each believes that nothing can escape him. Let those who like the European horse have their own way. I love the pure saddle-horse of the Arab. He is not part of a horse—somewhat of a monstrosity, and somewhat of a weakling—but a full horse, a well-balanced, a fit thing! In him the perfection of each limb, sinew, or muscle, though unmatched, still yield in value, as in rarity, to the symmetry and congruity of the whole structure. This admirable fusion of each separate part, this unity of the whole, is not, as it were, eclipsed, overshadowed, or humbled, by any too salient good point, neither is it disgraced, nor disfigured, by any flaw or meanness in his combination. In him, each proportionate perfection is hardly equalled by the one special usurping excellence of any other breed. Examine him as you please, and where is he found wanting? His head—what equals it in beauty, much less in expression? What steed has his development of brain? Where else so full, so dark, so lustrous an eye? Where such courage, such intelligence? His middle piece—where, together with slanting shoulders, and sufficient neck, such rounded ribs, such room for heart and lungs, such a strong, arched croupe? His chest! How ample, how nervous! His legs! What others are so powerful, so sound, so lasting? Where is the sinew so devoid of flesh and fat, clear from the shank, so cleanly strung? Examine the texture of his bone, close-grained, heavy, with diminished orifice; as compared to that of any other horse, it is like oak beside deal. Descend to his hoof, and you find it arched and open, tough, elastic, and sound.

When put to the test, his many virtues more than realise what his appearance promises. In the speed of the turf he is not an unworthy antagonist to the horse whose speciality it is. At long distances, a day's gallop, *facile princeps*, none can be found to enter against him. In sure-footedness, the mountain pony is thrown into the shade beside him —as the daily drudge he stands alone—he can slave always—his abstemiousness is not surpassed by the ass. In longevity and duration of service he is classed with the mule. In docility, in spirit, in sagacity, in attachment to his rider, no one pretends that he has a rival. He is the perfect horse—the horse of most pure blood : in him there is nothing excessive, salient, nor exaggerated : he is a full horse, in nothing mean or little : in no part wanting, in nothing exaggerated. Of other horses it is said— he is hardy, but ugly—fast, but delicate—strong, but without wind—a good thriver, but faint in the sun —pleasant, but unsound. The Arab horse alone needs no qualification : he has no bane : there are no "buts" in his character. There is nothing about him which one would enlarge or prune : all is congruous, sound, well-adapted, fitting, sightly, harmonious, polished, and beautiful, in his mechanism. No one feature about him is common or vulgar : he bears the stamp of nobility, not less in each separate point than in his whole frame. There is no coveted quality which cannot be got in perfection from his blood, by consecutive selections, to the required end: he is the fountain head of every saddle-horse good quality. *He is a pure saddle-horse.*

What blood runs in the veins of yon steeds who have just passed the winning post, amidst the shouts and cheers of the throng? The unmixed blood of the Arab. Whence comes yon slashing trotter of fabulous price, that rushes along the hard road between the shafts, whirling the New-Yorker to his country seat amongst the hills—where dwelt his ancestors? On the rocky hills of Araby. What is yon docile, easy-paced Naraganset, caressed by fair hands? What is yon steed, that, amongst the snows of the rocky Caucasus, bears his rider clear from the Russian sabre? What is he that, on the sultry plains of India, faces the wild boar and takes off the spear of honour? What is the nag which the veteran on leaving the East bequeaths with a tear to his friend, as something which cannot be bartered for money, endeared as he is by a thousand recollections of pleasure and service—what are they all, but descendants, more or less direct, from the pure-blooded desert saddle-horse?

To conclude: a breed of saddle-horses cannot be released in the attainment of perfection from those conditions which nature has affixed as its price to the whole genus. A body of good saddle-horses should carry their riders some eighty miles on five consecutive days, rest three days, and return at the same speed. To such excellence, however, there is no short cut. What we would have, that we must pay for. The price of perfection is purity. High breeding and pure blood are the result of selection, and not of crossing. Perfection cannot be bettered—to purity an addition is a stain.

I advocate in practice what we all admit as our principle, that "*like begets like, or the likeness of some ancestors.*"

SELECTION OF SIRES.

"Choisissez l'etalon et choisissez-le encore."—*Proverbe Arabe.*

Whilst purity of blood in sire is a *sine quâ non*, it is not the only requisite. A horse may be quite pure and utterly degenerate. Two horses of exactly the same blood, own brothers, may be as dissimilar as sires as in performance. Hence the horsebreeder must be as rigid in his requirement of what usually is found in a good sire, of personal health soundness and stoutness, as he is of pedigree. On this subject their remain a few facts to be noticed. The reader will remark, that there are many results in breeding quite beyond the control of the breeder; for, though good horses can come of good blood only, yet it is clear that all so bred will not necessarily be deserving that character; that in fact there will be excellence unaccountably bequeathed in widely different degrees. So that, though the sire A gets the colt B out of the mare X, and B turns out superexcellent, it is not at all certain that the colt Z from the same sire and dam will possess the same characteristics in the same degree. Still less, and this is important to consider, is it at all to be calculated on, that if even the colts B and Z do exhibit exactly equal virtues, that therefore each will in his turn and with equal opportunity be alike fortunate in bequeathing in full his own good qualities to his offspring. Thus it has been known, amongst Race-horses of the highest breeding in that kind, that some stallions have shone on the Turf more than in the Stud, and that others have been more celebrated as the getters of winners than as themselves horses of great speed.

In the mare the difference, in certain individuals, between their own performance and that of their offspring is still more decided and marked. Some mares, from their excellent qualities as dams, have thrown winners to any sire. And with reference to this I may notice that mares which are eminent in the stud, throw after the sire, whilst such as throw after themselves, produce stock usually inferior both to themselves and the sire.

Such being the case, it is necessary that the breeder should qualify himself, as far as possible, to discriminate between a good performer and a good progenitor; a mare that performs well, and one fitted to throw good performers. On this head it is exceedingly difficult to give specific directions, nor do I find that the subject has been treated of by any author with whom I am acquainted. With regard however, both to sires and dams, the first requisites (after purity of blood) which should be insisted on are, without doubt, soundness and vigorous health. If such are wanting, it will probably happen that the offspring will be less able than its progenitors. As regards the sire, he should bear well-developed those several points which distinguish him from the gelding and the mare : whilst the mare should not less distinctly possess those female shapes and characteristics which are proper to her sex. A mare whose figure leaves it for a moment to be doubted whether she be mare or gelding will usually throw foals inferior to herself, she may even be barren ; whilst the sire that might be mistaken for a gelding will seldom earn much fame in the stud. In a perfect animal these distinctions of sex are noticeable throughout his whole make.

This, however, is one of the minutiæ of horse-breeding, and there remains to be pointed out to stud-masters matters of greater import.

I am of opinion that one principal cause of the inferiority of our own and other saddle-horses arises, irrespectively of their bad pedigree, from the erroneous system on which their sires are selected, for we habitually choose rather from the presence of some special good quality, than as we should *from the absence of faults*, and the general harmony and perfection of the animal in his whole conformation. Thus, we are prone to use a sire if he be eminent in one good quality, though he be below par in other respects. We choose one because he can gallop, another because his point is fencing, or because he is a good thriver, satisfied with this partial excellence, and overlooking, for its sake, the want of soundness, the infirm hoof, the evil temper, the want of wind, or some other deficiency which should have damned the animal. One of the results of this has been that the best of our horses are deficient in evenness of character, and bear about them, with some excellent points, others that are very detrimental, the reverse of that perfection of the whole frame and finish which I so value in the Arab, which enables him to do all or anything that his rider may require of him.

"The Arab horse," says Abd-el-Kader truly, "is the result of the necessities of his rider;" on his possessing, in at least a moderate degree, all the great requisites to perfection in a horse, may any day depend the life of his rider." The exaggerated development of any one good quality cannot compensate to him for any flagrant deficiency. For instance, speed, so transcendent as to enable an animal to overtake all fugitives, or to baffle all pursuit, would be useless to the owner if it were accompanied by unsound legs,

L

a flat hoof, an over-voracious appetite which the owner could not fully supply, want of stoutness, non-endurance of thirst, or any other radical defect; because, though speed might be all-important on one occasion, at the next moment the life of the rider might equally depend on the soundness, the stoutness, or the endurance, or even the temper of his horse.

Thus, in the selection of sires, I hold it to be necessary to make sure that there exists no radical defect in the animal chosen, rather than to look about for one famous in some points, but grossly deficient or defective in others.

ON CLIMATE.

"The Arabs declare that the air of Yemen causes a degeneracy (in horses) in the first generation."—*Burton.*

"La race chevaline, telle qu'elle existe aujourd'hui en Afrique offre un hereux melange de tons les dons qui sont l'apanage du cheval dans les pays de vastes espaces et d'ardent soleil."—*Daumas.*

I am not aware that any author has treated of the influence of climate on the horse. A few passing unconnected remarks, let fall here and there, and these often of a contradictory nature, are all that I have met with on the subject; and yet I have no hesitation in saying, that there is nothing in the matter of horse-breeding, with the exception of 'blood,' which so imperatively calls for our attention.

Stonehenge alludes cursorily to the subject o climate on two or three occasions. Thus, page 19, he says: "It is probable that much of the wiryness of leg and lightness of frame in the Arab is due to the sandy soil in which the grasses of these oases take root.

Besides this, the *dry air* may have something to do with the development of muscle and tendon," &c. : and at page 55, speaking of the English horse, he says, " Much of this excellence is due to the *climate* and soil of the country, which encourage the growth of those fine grasses that exactly suit the delicate stomach of this animal." Thus, we must suppose that the very dissimilar climates of England and of Arabia are both particularly well suited to the production of the horse ! In a happier mood, however, the same author speaking of stables, tells us, page 187 : " The two most important points to be regarded in the choice of a situation are, first, the power of excluding damp ; and secondly, the best means of keeping up a tolerably even temperature in summer and winter . . . . It should not, however, be forgotten, that the horse is a native of a dry country, and cannot be kept in health in a damp situation either in-doors or out. Nothing, except starvation, tells injuriously so soon upon the horse as damp ; when exposed to it he loses all life and spirit, work soon tires him," &c. In this doctrine I quite agree with Stonehenge ; I believe that nothing is so injurious to the horse as damp. But England has a damp climate, and England has not produced the worst of horses. Are the good qualities of her horses then attributable to her climate or to other circumstances ? I have heard it said that the grapes grown in the green-houses of England are superior to the best that hang from the vines in the open air in the natural climates of the grape. It may be so; but would this result, realized by skill, won from nature as it were at much cost and to a very limited extent, justify the assertion that England enjoys a climate particularly adapted to vine-growing? And can the annual production of a few horses, of foreign blood, nursed with great care and at much expense (even allowing that some of these are unequalled elsewhere, which I am far from doing), entitle the climate of their birth to be considered excellent for the purpose of horse-breeding, when it is found on inquiry that the whole process of their rearing has been a constant struggle to nullify the effects of that climate? Or must we not rather regard them as green-house plants; a triumph of art, and not certianly the result of a favoured climate?

General Daumas says still less about climate than Stonehenge. I find, however, in his work the following remarks. He is speaking of transferring the Barb to France. "The sun of Pompadour and Limousin is certainly not such as is needed by the natives of a hot country in the delicate years of their youth." Again, "The equine race, such as it is at present found in Africa, presents a happy combination of all those gifts which are the heritage of the horse in countries of vast extent and under a burning sun. These noble animals will not, however, come to meet us on the sea-shore; we must seek them in the *interior* and often far back."

"Prefer," says the Arab proverb, "the mountain to the plain-bred horse, and the latter to the *swamp-bred* horse, which is only fit to carry a pack." These opinions *à mon idée* are quite correct as far as they go.

My first experience of the horse was in Tasmania and Australia; there, on his back, through many a ride, in heat and cold, by daylight and starlight, on the road and in the bush, broken, half-broken, and unbroken; grass-fed and stable-fed, well-bred and ill-bred—I first made his intimate acquaintance. Living amongst Englishmen, it is not wonderful that I should have listened to their encomiums on the horse of their father-land; of how tall, how fleet, how beautiful he was; and, as a consequence, become a true believer in the merits of that animal. Whether I heard or read of him it was as the first of his kind. The fact appeared as incontestable as that Rome stands by the Tiber, and, unquestioning, I believed one as firmly as the other. It was whilst travelling in Europe and Asia that my orthodoxy on this point became at first shaken and at last upset. I was in search of no information concerning the horse, but in riding or driving through various countries, I could not, from my former habit of the saddle, fail to remark, that which so soon attracts the practised horseman, the qualities of the beast I rode or drove. It first struck me that there were certain general characteristics noticeable in certain neighbourhoods or districts: in some, the horses were all more or less stout; in others, the reverse. In some, their performances were better than their figures seemed to promise; in others, less.

In coaching through the swampy country between Rome and Terracina, looking at the dank ground and canal-like drains full of fœtid water by the roadside, I interrogated the coachman concerning the health of the inhabitants of the district: he informed me that the locality was so unhealthy that neither the men employed there nor the horses on the road could stand the climate for more than two or three years. This incident was what first turned my attention to the effects of climate on the horse. As I proceeded on my wanderings, I had excellent opportunities for making observations on the subject, as well as an accumulation of instances already witnessed to refer to, which at once presented themselves to my mind. The results of my observations, as well as what I have been able to glean from other travellers on the subject, are as follows: that the horse, though of a very accommodating constitution, and of a very wide habitat, like all other animals is very materially influenced by the climate in which he exists. That a suitable one is as material to his perfection as is blood or food. *That the horse most excels, in proportion to his size and figure, in a hot and dry climate.* That he is frequently, but not necessarily, nor always, diminutive in such climates; *where alone he reaches the full perfection of which his nature is capable.* That his utility is various in various localities, and corresponds to the fitness of the climate in which he lives. That in the hot and moist climate he sinks to the deepest debasement. That it is a fact—and one easy of verification—that every race of horses, without exception, which has acquired any celebrity in the saddle, is either bred in a hot and dry climate, or traces its celebrity to strains of blood originally derived from such climates. So also the horses of all countries exhibit in their qualities a marked relation to the humidity or dryness of the air in which they have been reared.

Thus it appears, too, in support of what I have advanced, that you cannot breed as bad a horse, inch for inch, in the Sahara as on the fenny banks of the Hoogley. The dry climate of the Sahara invigorates the first, the damp climate at the mouth of the Ganges undermines and impoverishes the constitution of the other. The first will perform much more than the Englishman would anticipate from his figure; the other, far less. In the one climate he is stout, abstemious, mettlesome, and hardy, in proportion to his figure ; in the other, he is peevish, soft, washy, relaxed, and unenduring, even in spite of a good figure. In the latter, a good figure is difficult of attainment ; in the former, a bad figure is the exception. In the dry climate, he agreeably surprises ; in the damp, disappoints. In whatever country I have been, in the old world or in the new, in the northern or the southern hemisphere, these results have forced themselves on my observation. Australia, as I have before noticed, is an exemplification of this remark.

The English thoroughbred horse is a tall fine-looking animal, and unrivalled in speed of his sort. Whatever excellence he possesses, he owes to his Eastern blood, to the blood from an arid climate. West Australian was a great horse ; I saw him win the Derby ; but had he been bred in the desert, he would, if my theory be correct, have been a faster horse, a stouter horse, and a more valuable sire. It may be remarked, that no horse of his size ever came from the desert. Let us postpone that consideration to the next chapter. I speak now of the performances of given figures in various climates, rather than of their production. The English-reared horse, that would carry you up to hounds for forty minutes in Leicestershire, would have performed the same service for one hundred and twenty minutes had he been bred in Nejyd. The English horse exiled to India, becomes ailing, enervated, diseased, and useless, from the effects of the humid atmosphere. His importation there has ceased. Taken to Australia, he enjoys at once a measure of health, and becomes robust to a degree to which in England he was a stranger. He immediately improves in the dry, and deteriorates in the damp climate. The Arab taken to India, whilst losing by the change, outshines, from his great stamina, all imported horses, and even those native to the country.

The horse of Australia is another instance of the correctness of my statement; it is in the arid inland districts that he shines, in proportion to his figure. The difference between two horses equal in appearance, one of which has been reared on the coast and the other inland, is both very great and well known; and it is in the hot, damp, agueish districts on the sea-board of Queensland, where he has been found most to deteriorate; and it is further precisely at the season of the midsummer tropical rains, that his health there is found to be most feeble and his powers of endurance the lowest.

Man has been brought to exist in all the climates of the globe,—in some he retains or reaches the maximum of his strength; in others he dwindles and deteriorates. We all know how susceptible he is to the influences of climate. So powerful, indeed, is the influence of climate on every living thing, that its effects can hardly be over-estimated. Examples of this truth are numerous. The sheep, for instance, taken from one country, becomes larger or smaller, —bears more wool or less, which becomes finer or coarser according to the nature of the climate to which he is removed. The dog experiences similar changes. Vegetables and fruits of every sort are subject to the same law. If you have a vine in England, which produces only acrid little berries unfit for food, you have but to raise a glass-house over it and create within the climate of Andalusia, and you will grow a luscious fruit, from which you may express the juice and manufacture Mansinilla or Amontillado.

Man himself is notoriously subject to the same influence. Thus, the Anglo-American, with his European blood, food, education, and habits, has already begun to show the tendency to the beardless chin and hatchet face of his predecessor the Red Indian. So also families resident a few generations in the West Indies, have constantly and strikingly exhibited an approximation to the type of the aboriginal of that climate. The case with the horse, if not identical, is similar.

I have before remarked, that the most abstemious,

vigorous, swift, and enduring horses are found between the 20th and 40th degrees of north latitude, or thereabouts; and from as far back as we can trace the animal, he seems, as well as now, to have flourished in the same localities. So when Hannibal led his Africans into Italy, the horses from "Numedia's burning sands," did not fail to prove their virtue at Cannæ and many another well-fought field. So formerly, as now, Persia and Asia Minor were famous for their cavalry horses; and all these are dry and hot countries.

Now, in hot and humid climates, as in the vicinity of Rio de Janeiro, for instance, the reverse of all this is the case. Close on the southern tropic, this city is situated in a moist, pluvious district, where the deep verdure of the trees, the superabundant vegetation, the yellow-fever, the hazy atmosphere, and the lethargic habit of the inhabitants, alike bear testimony to the excessive humidity of the climate. Here we find the horse, as well as the sheep and the ox, small, ill-conditioned, and inferior. On one occasion I saw three or four cavalry regiments collected near that city. In the whole assemblage of horses, I doubt if there was one that stood over thirteen hands high. In fact, it was the ludicrous sight of cavalry soldiers mounted on ponies.

From a long and attentive consideration, as well

as from considerable experience on the subject, I have come to consider climates suited to the horse in the following order :—

1st.—Dry and hot.
2nd.—Dry and temperate.
3rd.—Dry and cold.
4th.—Damp and temperate.
5th.—Damp and cold.
6th.—Damp and hot.

It is on these premises that I have formed my high estimate of the horse-breeding capabilities of inland and central Australia. The probabilities of sparse population, pastoral pursuits, and infrequence of rail-roads in many of our arid central countries, must also be considered another guarantee for the future good qualities of our horses.

It is on this estimate of climate that I account for the softness and delicacy of the English horse. His future will, I believe (regard had to selection) be in proportion to the Eastern blood employed in the stud.

In this country we possess the great elements of the amelioration of the horse, climate and food that are admirably suited to his constitution. Starting with the miserable beasts we now have, I believe that, in the course of ages, by constant and judicious selection alone, and without the introduction of pure blood, a breed of saddle-horses of much virtue and distinction might be created. The use of pure saddle-horse sires would, of course, much facilitate the undertaking, and will, I dare say, some day be had recourse to.

FOOD.

"Si nous n'avions pas vu que les chevaux proviennent des chevaux, nous aurions dit : c'est l'orge qui les enfante."

Sidi-Hamed-ben-Yousseuf.

A full supply of good food, is, of course, essential to the perfection of the horse. This, however, does not interfere with the cultivation of abstemiousness, a virtue quite unknown in our stables, where gluttony is more prized than *sobriété*.

If the quantity of food supplied to the horse is habitually insufficient, or still worse, if the quality is inferior, much less unwholesome, bad results quickly become apparent. The consequences, however, of insufficient food, and of inferior forage, are very dissimilar. If the quantity of food for several generations be somewhat insufficient, but of good quality, the size and stature of the animal becomes curtailed, whilst his figure will loose none of its proportions, and his stamina will correspond to his make, and be even increased. If the quality of the food be very bad, the animal will suffer proportionately both in size and stamina; but if the food, as grass, be merely somewhat too soft, too sour, or wanting in nutriment, the animal, whilst losing but little in bulk, will be found to be most seriously deficient in stamina. Hence, a better horse is grown on a limited dietary of good food, than on a full allowance of indifferent forage. " Between a racehorse reared on corn, and another confined to hay and grass," says Stonehenge, " the difference in value would be 1000 per cent.; and in first-class hunters, though not so great as this, it would be very considerable."

The effects of the quality of food allowed to the horse will be apparent in his power of exertion, as compared to his size, figure, and blood, in every climate, and under all circumstances; but the quantities of food necessary to him vary according to climate. He is most abstemious where he is most perfect, in a dry and hot climate. He is most voracious where the climate is damp and temperate.

No Australian who has travelled far on horseback can have failed to have noticed that his horse requires, and eats a much greater quantity of grass, in a timbered, damp, and cold country, than he does in a hot and dry one, which arises from the increased necessity of food, resulting from the effects of climate, as well as from the actual inferiority of the grass found in such climates. Thus, an English horse will devour twice the weight in rye-grass and clover that an Australian horse will consume on the plains of the Campaspe, Loddon, or Lachlan ; and yet, whilst the first, as shown by Stonehenge, will barely live, the other will work hard. Climate, as well as food, it is true, has its influence here. But in contiguous enclosures, where the climate of course is one, but, as is frequently the case, the pastures different, the truth of what I state is very noticeable. One reason, perhaps, of the superior quality of grass in hot and dry countries is this : vegetation takes place rapidly in the short springs of such climates, bringing the grass and seed quickly to maturity ; the spring then finishes, as it were, in a day ; the dry, hot weather arrives, at once converting the grass, which is at its prime and has had no time to deteriorate, into fields of sweet-scented standing hay ; the virtue of which is gradually, but seldom entirely lost, until the advent of the autumn rains, which usher in the fresh winter grasses. I believe firmly, however, that the intense light of the sun is also favourable to the quality of vegetation, in a greater degree even than is usually admitted, and that the nutritive elements, not only of grass, but of oats, barley, Indian corn, &c., grown in dry climates, are more perfect than are to be found in the same cereals were such light is absent.

The corn preferred by the Arabs to all others is barley. I have already noticed that twelve double handsful of this cereal daily is found to be abundant feed for their horses.

I have heard it maintained, in conversation, that the low stature of the Arab and Barb is attributed to the dryness of the countries in which they are found, but there are several facts which clearly militate against this idea. The Australian horse, from the driest districts for instance, is only smaller than that of England, when very good reasons for his falling off in size can be adduced, independent of climate. The low stature of Arab and Barb I attribute chiefly, if not entirely, to the sparing supply of food on which he is reared; and to which his ancestors have been restricted for many generations. Grass is scarce amongst the wandering Arabs, and even when such is not the case, the value of the animal and the fear of the *gazzou*, or raid, confine him to the vicinity of the tent, and allow but a partial use of what does exist. Neither is grain cultivated amongst them, but sparingly received in barter or exacted as tribute. Hence they are put to many shifts for horse-feed, using milk, dates, &c., as most accessible in their circumstances. These, it is true, they now prefer to grass or hay, but perhaps without any solid reason for so doing. Besides this, well-grown horses are in fact to be met with frequently in the arid climates where forage is more abundant. I once travelled, for instance, with a caravan of seventy-two Eastern, not Bedouin, horses belonging to Abbas Pacha, then Viceroy of Egypt. They were from the neighbourhood of Damascus, and each about sixteen hands high. Stonehenge, in speaking of the Turkish horse, says that he " seems merely the Arab *developed by higher food* into a larger size, and more massive proportions. The horses of Constantinople are often sixteen hands in height, and with very elegant proportions, and a crupper more highly developed than that of the Arab." This bears out my experience in that city, having often seen pachas, &c., in the street at a walk, mounted on magnificent bay horses exceeding even sixteen hands in height, attended by servants walking before, beside, and behind them; one holding by the cantle of the saddle at each side.

By some, again, it has been supposed that the Arab horse is a small species of horse; that his diminutive size, more frequently under than over fourteen hands, is not the result of the sparing diet to which for ages he has been subjected, but is a natural characterestic of a distinct and separate variety. This position is, I certainly think, not tenable. First, it would suppose him always to have been of unmixed blood, which no one claims for him; on which subject let me refer my reader to Youatt on the Horse. Then we find the Arab horse himself varying in size in different localities, and always bearing in this respect an intimate relation to the quality and measure of his diet. But it is, as I have before stated, in England that this has been fully set at rest, where the results of the full and even forcing diet have gradually but surely become apparent. "Out of 130 winners," says Stonehenge, "in the middle of the last century (since when the thoroughbred horse may be held as uncrossed) only eighteen were fifteen hands and upwards; whereas now, a winner below that height is a very great rarity indeed, even amongst the mares. . . . The average at present may be fixed at fifteen hands three inches," &c.

It is, however, very doubtful if nature has not, except in rare instances, restricted the full powers of saddle-horses for general purposes to a frame very much smaller than that to which its growth is so often forced amongst Englishmen. In Australia, moderately heavy men find themselves better carried on long journeys by a rather large than a small horse. Where however, as here, horses are almost all equally indifferent in blood and figure, the larger one will be found preferable. But this is no reason why a very much smaller horse, more carefully bred, should not surpass either. Equally bred, and of equal figure, the larger will of course always be preferable. Perfection of figure, however, is more rare in the large than the medium sized horse. "Buy a large horse," say the Arabs (who have not what we should call a large horse), "and barley will make him go."*

In treating of the horse of all countries, as do Stonehenge and Youatt, but few maxims for horse-feeding can be adduced as universally applicable. The subject of food is so mixed up, intimate, and interlaced with that of the effects of climate, that the two considerations become at every step merged and inseparable. As with the human race, in obedience to the laws of animal chemistry, it is known that a regimen, excellent in some situations and circumstances, is and must be detrimental, excessive, or inadequate in others; as likewise, the grass that is amply nutricious in one climate is absolutely insufficient in another; so no specific result, either as regards health, size, speed, or vigour, is produced, which is not attributable in strictness, in part at least, to the joint rather than the separate action of these causes; since the very quality of every food, as well as its adaptability, is one of the primary consequences of climate.

---

* "Cherche-le large et achète
L'orge le fera courir."
Abd-el-Kader.

Almost all that we meet with of variety in the physique of the horse, spread over the face of the earth, whether it regard size, shape, or quality, has resulted from the joint and united action of selection (or its absence), climate, and food.

RECAPITULATION.

"Mark me, Jack."—*Poins*.

From the facts placed before the reader in the foregoing pages I draw the following conclusions :—

1st. The saddle-horse has occasionally been bred with considerable success in various climates.

2nd. A fitting climate is of the utmost importance to the development of the full powers, stamina, and utility of the saddle-horse. Certain valuable characteristics in full perfection in him are coexistent with certain climates only. The stoutest saddle-horses of which we have any reliable records, or traditions, appear heretofore to have existed and to be still found between the 20th and 40th degrees of latitude North, or thereabouts. Such breeds of saddle-horses as have been famous outside these limits have invariably traced their excellencies to sires from within the said latitudes.

3rd. The mental and physical qualities, both good and bad, of the horse are susceptible of cultivation, and transmissible to offspring; hence the impropriety of breeding from unsound or unfit sires, which is embodied in the axiom, that "*Like begets like, or the likeness of some ancestor.*" On fitting climate, food, and the *strict application* of this axiom, must always depend the success of the horse-breeder.

4th. The Arab or Eastern horse is a saddle-horse, bred after the above axiom.

5th. The Arab horse is a pure saddle-horse, and not a racer. His blood is pure, and possesses the requisites to this title, as laid down by Stonehenge, page 82, and already quoted.

6th. On the same principle, no horse of any breed is so pure as the Arab.

7th. The thoroughbred of England is almost entirely of Arab blood. So small has been the foreign admixture that, practically, his blood may be called Arab blood. The differences now noticeable between the thoroughbred and Arab are easily accounted for, and are to be attributed to *selection, climate, and food*. Selection, by the choice of mere fast horses. Climate, by one moist, and encouraging growth. Food, by being pampered, in place of being judiciously stinted.

8th. The Arabs have never bred horses for racing

alone. Hence, when the English horse races against the Arab he puts to the test the only quality in which he himself excels, and is proclaimed the victor, and the best horse, leaving untried several more necessary excellencies—all those points which constitute, in fact, a good saddle-horse — which it is well known he does not possess, but are all preeminently the characteristics of his antagonist.

9th. Health, vigor of constitution, abstemiousness, stoutness, soundness, and longevity, are points in which the Arab horse is without a rival. "He presents," as Youatt says, "the true combination of speed and bottom."

10th. The saddle-horse of England is bred contrary to the theory laid down for breeding by Stonehenge and other writers, and contrary to that practice by which alone the best saddle-horses have been obtained elsewhere.

11th. Englishmen claim for their saddle-horses the merit of surpassing all others. I can find records of individual English horses having greatly distinguished themselves. I can find no instance of their ever having done so in a body. Where they have been collected in bodies, as in the cavalry service, all testimonies seem to agree, that their capabilities for work are much below those of Eastern horses. I can meet with no writer who has had personal experience of each who does not pronounce in favour of the Eastern.

12th. The management and performances of Eastern horses are but little known in England, and seldom believed when recounted : the usual reply of Englishmen being, "Our horses are the best in the world; they cannot do these things : therefore," &c., &c.

13th. The sums of money expended on horse-breeding, and trying to perfect the horse, in England have been enormous.

14th. The saddle-horse of England is soft, unsound, and a failure.

15th. Racing in England, so far from being the means of improving the saddle-horse, has been his destruction. This fact is acknowledged and regretted by several English writers, who loudly claim Government interference on the subject.

16th. The saddle-horse can never be brought to perfection until he be bred pure, as in the East.

17th. The soundest, stoutest, and purest of horses is the Arab horse. He is a pure saddle-horse, unequalled in his performances, and unrivalled in his desirability for this purpose. No saddle-horse is excellent, except through his blood.

PART III.

"Dehors, les étrangers, dehors !
Laissez les fleurs de nos prairies,
Aux abeilles de notre pays.
Dehors, les étrangers, dehors !"

SADDLE-HORSE BREEDING IN AUSTRALIA.

"Je veux un cheval docile
Qui aime à manger son mors
Qui soit familier avec les voyages
Qui sache supporter la faim
Et qui fasse dans un jour
La marche de cinq jours."
Le Chambi.

" Nor is man's influence over many of the animals less marked. The habits which he imparts to the parents, become *nature*, in his behalf, in their offspring."—*Testimony of the Rocks.*

In the first part of this work I have laid before my reader, a collection of facts, which I hope will be suggestive and useful to the breeder of saddle-horses ; in the second part, I have made deductions and laid down certain principles, which appear to me to be based on those facts, and I will now proceed, in conclusion, to show how it seems to me that those principles should be reduced to practice.

Hitherto in Australia, breeders have neither bred pure nor, as a consequence, good saddle-horses, nor brought to market at a remunerative figure such as they have bred. This failure, as an investment, has resulted in great measure from the want of management which has always attended such speculations.

In all countries there are situations more favorable than others to the production of the horse. In Victoria, particularly, there are few sites for this pursuit which are perfectly unobjectionable, and unite every requirement, and yet the selection of a fitting field for this undertaking is quite as important as is the site of his vineyard to a vine-grower. Both vines and horses may be produced almost anywhere, but the quality of the produce will be in both cases widely different in different spots. New South Wales offers greater facilities for this undertaking than Victoria, and nearer the centre of this continent the requisites will be found both more general and superior in degree.

Could I have a site and all facilities for saddle-horse breeding in Australia, just cut out to my fancy, I would have it in this way: On a flat of poorish, salty soil, I would grow oaten and barley hay. This for nine months out of the twelve, should be the food of the stud. Between the hay-racks, where the horses were fed and the watering-place, I would have a mountain, bare, rugged, rocky, and steep; over this the horses should travel daily to water, and back again to their feed. The distance to be accomplished daily would depend in a great measure on the steepness of the mountain, and be graduated besides to the various ages of the horses, ranging, say, between ten miles a day for the four year old horses, and two miles a day for those under a year. During the three spring months of the year, I would indulge the stud with abundance of luxuriant grass and salty herbage, water easily accessible, and almost a complete cessation from toil. The site of my breeding-ground should be in the most favored path of the hot wind, and in the most arid district that could be found.

From the importance of a fitting site for this industry, united with other causes, it will be found more easy and profitable to breed good horses in herds, in chosen spots, than in ones or twos all over the country. Breeding saddle-horses, on a large scale has been frequently undertaken in Australia, and with such unprofitable results as to be at last considered an all but ruinous undertaking. This, however, is not surprising, considering the bad management which has generally been pursued, to which cause entirely I attribute both its unsatisfactory results and its unpopularity; to which must, indeed, in some degree, be added, the large outlay first needed, and the necessity of four years passing over before any return for the outlay can be realized.

In Victoria, at this day, really superior colts weight carriers, handsome, of high-breeding, and unbroken, would readily bring £15 in the market; while such colts, at four years old, and perfectly broken to saddle and harness, would not be worth less than £30 a-head. Were the average value struck of such horses during the past twenty years, I believe it would show not less than £12 for the colt, and £25 for the broken and made horse. The average value of five year old fat cattle, of both sexes, is now about £6 a-head, and it has not certainly exceeded that figure as an average during the same period, and yet cattle have been a favorite investment, and horses almost always a very unremunerative one, attributable entirely, I believe, to the bad style of management which I have already described in my chapter on the Australian horse.

The first point to be considered by the horse-breeder, who seeks quality as well as quantity, is the site of his breeding run. Its climate must be his first care. This should be hot and above all dry, and remote from the sea-board and the influence of the ocean. A first-rate saddle-horse cannot be bred in a damp atmosphere, or on a pasture too luxuriant and fat. Besides this, the ground should be hilly, and in places steep and rocky, that the muscles of the animal reared on it may become enduring and well-developed, his frame nervous and tolerant of labor;

that his feet may be arched, hard, and tough, and the useful habit of picking his steps and minding where he places his feet early engendered. " *Préfère le cheval de montagne au cheval de plaine, et celui-ce au cheval de marais, qui n'est bon qu'a porter le bât.*"

Various defects are noticeable from breeding horses in improper sites. As an instance, the horses from the soft plains of the Lower Namoi, N.S.W. have come to have feet of about the size and shape of a soup-plate. Hence, when hilly and in part rugged ground is not attainable, a hard soil should at least be insisted on. Another feature in the choice of a horse station should also be kept in view. Though the grasses which grow on flat, rich, or damp soils, always produce an inferior horse, yet to be used as a change, and in some cases medicinely, and as an occasional spur to the condition of the animal, some portion of such country, where attainable, must prove a valuable adjunct to the poorer country already described; at the same time, it is by no means indispensable. It may also be added, that the more denuded of trees generally, and the freer the action of the sun, the better will be the country for this purpose; a few patches of timber only being necessary for shade and shelter.

In Victoria, I know of no hilly country which is sufficiently dry and hot and removed from the influence of the sea air, and at the same time of a saline quality, and producing a hard and tenacious grass rather inclining to a tusic than a sward, so that in this colony, for this purpose I esteem most the sound and hard portions of the plain district lying between the rivers Campaspe, Murray, and Richardson, and it is on this country that I have always found the stoutest horses of Victoria in *proportion to their figure*, surpassing horses of a better figure bred on the rich soils of the sea-board.

In this respect New South Wales offers greater facilities than Victoria, and I know many admirable sites in that colony, but none perhaps superior to the station known as Uabba on the river Lachlan. Here you have a dry and scorching climate, hotwinds fierce and continuous, which leave nothing to desire ; the steep rocky Uraral Mountains substantially grassed, with an ample supply of hard plains at their feet. From this country 300 excellent horses might be turned off annually.

The horse-breeder, having chosen a spot suitable for his purpose, his first steps should be to fence securely and subdivide it. Without this nothing can be successfully undertaken, as the horse roaming at large becomes wild; resignation, that quality so prized by Abd-el-Kader, and so necessary in itself, becomes entirely a stranger to his composition ; the foal and yearling get galloped to death to keep up with the herd, and frequent manipulation of the young stock becomes impossible. Accidents are also proportionably more frequent amongst large than small bodies of horses, for which reason I think not more than fifty should be kept in one paddock.

In the selection of brood-mares, where all are so indifferent as in Australia, there is unfortunately but small room for a choice. Were there a choice, I should say select to breed from just what you would select to ride. Let them, however, be as stout, sound, active, good-tempered, and pleasant as can be obtained. Of colour be particular: forswear roans, sorrells, duns, and pie-balds. Light-coloured horses of every hue, whether bays, chesnuts, browns, or greys, are soft and inferior. The Arabs, with reason, esteem deep colours only: if it be a bay, let it be blood-red, golden, or a brown bay ; if a chesnut, let it be of the darkest shade, which looks like black in the distance, and is perhaps the best of all colours, denoting the best constitution, and as being more commonly allied with speed, temper, and endurance. Size as well as colour is a matter in which the Australian breeder will have no difficulty in suiting himself; therefore let him start with large mares, where action and soundness are united ; with it one difficulty will be removed from the path of the regenerator, and tend to expedite results; which, choose what you like in this country, can only be gradual, and must be achieved principally through the sire.

The mares having been obtained, fit sires are still to be purchased, which, though difficult to be had, are still attainable. The requisites in the sire are, that he should be sound, a good saddle-horse, and pure of his sort—a pure saddle-horse, in the sense I have already explained. He must, in two words, fully to meet these requirements, be an Arab horse of high caste.

The mare should not be put to the horse earlier than three years off, and first foals are, I think, seldom equal to those that come after. As the proceeds of the first cross, which would be half-bred Arabs, became numerous enough to keep up the breeding stud to the requisite strength, the original mares would, of course, be all disposed of; and, if two sires were obtained, the get of the sire *A* be put to the sire *B*, and *vice versâ*. When the three-quarter-bred mares came to require the horse, fresh sires would of course become necessary, and the first ones would have to be disposed of.

When three crosses had been effected, I should recommend that the mares selected yearly for the stud should not be chosen with reference to pedigree alone, as heretofore, but with due reference to perfection of figure, size, temper, colour, and performance; giving, however, a preference to the strain of any stallion whose get exhibited superior excellence in quality as distinguished from appearance. Were this system pursued for twenty generations, on a sufficiently extensive scale, it is probable that the importation of Eastern horses might with advantage be relinquished, and the gradual perfection of the breed, thus far established, be for the future entrusted to *selection* alone.

Let me point out to the reader, that in the system which I am advocating, there is no talk of crossing, no heavy sires to give size, nor racers to engraft speed on their offspring. For speed I depend on the Eastern sire, who has all the speed that is compatible with the admixture of more useful qualities which can be found in horses of their size. Larger horses of the same breed would, undoubtedly, be faster and more powerful. A breed with such characteristics may be created, but does not yet exist. My aim is to create this superiority of size, but not by crossing, and at the sacrifice of quality. This must be the work of time, and a full supply of food. *The Arab horse, transplanted here or elsewhere, will just arrive at that stature and bulk which the food and climate is calculated to raise him to, and sustain him at.* The abundant food of Australia should quickly foster a larger horse than the meagre diet of the Arabs can produce. So that I am for the pure Arab blood, which would, when bred here for a few generations on liberal but not forcing diet, reach the size which nature has affixed as the limit of the perfect saddle-horse in our country, which I am inclined to believe, would, as a rule, be found to be about fifteen hands two inches.

Having got thus far, the next subject which I have to impress on the breeder is, the necessity of careful culling and selection. It has always appeared to me, that the breeder of animals, of whatever sort they may be, should distinctly understand and fix in his mind the end which he purposes to attain, and strive for this end with unvarying purpose. He must be seduced by no sort of excellence foreign to his proper object. Were I breeding saddle-horses, I would reject from my stud a mare unfit to produce saddle-horses, though she should prove unrivalled on the turf, as certainly as if she were the merest weed. There can, I believe, be no good stud, where a fixed object is not kept in view, and culling most jealously and unhesitatingly exercised with strict reference to such object.

But besides the rearing of horses of good blood on good food, on a proper soil, and in a suitable climate, there is another requisite to perfection which demands attention; a requisite which has heretofore been quite overlooked by Englishmen and Australians, I mean abstemiousness. Abstemiousness is not habitual starvation, nor does it mean a constant insufficiency of food. The Arab horses, it is true, are abstemious in the widest sense of the word, for they can suffer long from hunger and thirst without injury, and live habitually on an exceedingly small daily allowance of food and water. The perpetual necessities of their lives have rendered this double endurance constitutional and easy ; hardihood has followed the frequent but temporary abstinence, but diminished size has been the penalty of constant deprivation of adequate nutrition. Australia is a richer country than those of the wandering Arabs. Australians are heavier than Bedouins, or Moors ; we are richer also, and can furnish more forage, and so rear larger horses. But this is no reason why our horses should not be taught to endure a hot sun, scarcity of water, and an empty stomach. Privation, when not excessive, may, to a certain extent, be made familiar both to man and beast, not only without diminishing the stature, but with the best results. A horse to be excellent and stout, must work and suffer. He cannot be habitually pampered, and yet enduring " *Les plus grands ennemis du cheval*," says Abd-el-Kader, "*sont le repos et la graisse.*" Neither must the practice of *sobriété* begin late in life, for says the Emir—

> *Les leçons de l'enfance se gravent sur la pierre*
> *Les leçons de l'age mûr disparaissent comme les*
> *nids des viseaux.*

A judicious measure of abstinence, both from food and drink, will be found materially to invigorate the system without diminishing the size. The Arabs are not called on specially to teach their horses to hunger with patience, for their whole life is necessarily one of short allowance. In the matter of teaching them to bear thirst, however, it is different. "Amongst the desert tribes," says Daumas, "beginning from the month of August, they only allow their horses to drink once every other day for forty days. The same practice is pursued during the last twenty days of December and the first twenty days of January." To such as breed saddle-horses for use rather than appearance, I would recommend a similar course. Besides the hardening of the constitution, which is certainly a consequence of a moderate abstinence, another scarcely less important result follows such habitual privations; I mean *resignation*, a matter little spoken of amongst us, but in high esteem with the Easterns. A total absence of this quality is, however, very remarkable in its effects; a striking instance of which will be found in the fact already mentioned, that the wild horse, which, when broken in, will, if loosed, gallop fifty miles in attempting to regain his freedom, will frequently refuse to carry his rider five miles. It seems, indeed, as if he were unable to work; his muscular strength, overpowered and prostrated by the irritation and despondency he experiences in captivity. This tendency, in a modified form, is well known to the Easterns, on which subject I have already quoted Sid Hamed-ben-Mohamed-el-Mokrani, and proverbial sayings on this head are numerous in the desert.

A horse that has not occasionally thirsted from his youth, is of little use in many parts of Australia. Nothing, I am persuaded, is more useful than an early but moderately conducted initiation into those hardships, which, to be easily borne, should early become habitual. Let not the breeder fear, then, to keep his horses athirst in the blazing sun; let his lessons be gradual and progressive; let him early know hunger, and how to bear it. But whilst his trials are frequent and severe, beware that they be not excessive. Of this his growth and condition must be the indices. When you have tried him with suffering, regale him with plenty; let him feast as well as fast,—keep up his spirits and his condition. So he will become enfeoffed of resignation, endurance, and vigour; and a hundred times more patient, capable, and robust, than if he were reared in the enervating lap of luxury and abundance.

It would be impossible for me here to point out the degrees of abstinence which should be proper to all circumstances. The nature of the locality, the descent, the previous habits of the animal in question, and a hundred other circumstances, would have to be considered ; and, even knowing these, trial and experiment could alone definitively settle the matter. As a general rule, it may be said, perhaps, that the loss of condition consequent on suffering which is quickly recovered has not been excessive. But, if after his abstinence the horse remains low, and is some days before he even begins to make up his lost flesh and spirits, if his food is ill-digested and his coat loses its gloss, he may then be pronounced to have suffered too much.

Though, however, the saddle-horse should be both well-bred and judiciously fed and brought up, there will still remain to the Australian breeder a difficulty, which, if not overcome, must be fatal to his success. This difficulty is his breaking and preparation for market. Labour is so high in Australia, especially that of the horse-breaker, and our horses so intractable, that no breeder has heretofore attempted to prepare, say, eighty or a hundred colts annually for the market. But even such preparation of the colt, that is, the mere breaking, would but partially set the matter right. In Australia, to *make* a horse, that is to educate him, is, when his education cannot be perfected in the course and process of useful work, more expensive than his breeding,—the skilled labour is dearer than the raw material. The result of these circumstances has heretofore been, that the breaker has usually reaped much of the profit which should have fallen to the breeder; for the raw four-year-old of good size, figure, and breeding, which is worth, say £15 in the market, will certainly bring £35 when he has been made perfectly tractible and handy. The profit to the breaker will, of course, be the difference between these two sums, less the price of the labour in breaking and the forage furnished in the interval. Besides producing a good horse, then, he must be brought to market well and inexpensively broken, or the speculation cannot be a success. Heretofore, as I have already pointed out, it has been little better than a failure from this cause. Now for this evil, there is, I feel persuaded, a very efficient remedy,—indeed, a specific—I mean, a skilful application of Rarey's principles. Without attempting to embrace every part of this subject, it will, perhaps, be useful to point out the manner of the application of this system which I advocate.

As early as possible after a foal is dropped, and as frequently as is convenient, and before Rareyfying him, I should recommend that the foal be caught. For this purpose lead the mare into an inclosure, and whilst the foal is sucking, put a strap round his neck. Let him make two or three bounds away from his mother, and then stop him. If he is under a fortnight old, in twenty minutes he will allow himself to be handled as much as you choose, and will resent nothing as long as he is not hurt. Rub him all over, wrap a cloth round him, ring a bell beside him, &c.; repeat this every day for a week, and you will find that the lesson will never be forgotten. At six weeks old, catch, halter, and Rarefy him. When he is exhausted and falls, fasten him securely, then wave cloths or blankets of various colors about, behind, and over him, striking him gently with them until he has ceased to fear them. Stand with your legs astride of him, cracking your stock-whip over his head (taking care that it does not hurt him), until he is used to it, which will not occupy five minutes; discharge your revolver, lightly loaded, from the same position, letting him see the flash and smell the smoke, taking care that you do not approach the pistol so close to the ears as to injure that organ. When he has been put down three or four times, unbind him, rub him well all over, standing alternately on each side. You may then lead him, your assistant following with a whip, and making him keep up. I need not say be most sparing in its use, and after half an hour spent in this exercise, tie him up securely out of sight of his dam. Let him see that you are not holding him, returning to him, from time to time, to rub him and speak to him. When he has been tied up half an hour, lead him back to his dam, and let him go. Continue this practice daily, until the foal will lie down when you touch his leg, which should be on the third or fourth day. Put him subsequently through this exercise once in every month, till he is a year old.

At twelve months old, having strapped up his leg, put bells round his neck, wave blankets, fire a revolver beside him, &c.; put on a saddle, and let him hop about a moment with it; put on a crupper somewhat tight, and girth him tight; when he has ceased to resist these liberties, which will be in 10 or 15 minutes, put a child on his back, the lighter the better. Any black child who is small enough will do, they are to be preferred to white children as lighter and more precocious; danger there should be none. When he has sat on the animal a few minutes, rubbed, patted him, &c., assist him to mount and dismount several times on each side. This being done, and the child seated, loose the tied-up leg, lead the colt with a short rein, and tie him up at the accustomed spot. Continue this treatment for a week without intermission. This week concluded, let the yearling at large, repeating the same performance once in every month, till two years or thirty months old.

At from two years to thirty months old, mouth the young horse, mount a light weight on him, and ride him two hours a day for a fortnight at a walking pace, and he should then be, not of course a made horse, but quite broken in, needing only a little practice as a three year old to perfect his education.

The treatment which I here recommend I tried to a certain extent just after the publication of Rarey's book, operating on two or three foals, and the like number of yearlings and two year olds. Every one in the bush was then talking of Rarey, and I, like others, felt anxious to try the working of his theory, which I applied to all the unbroken stock I could lay hands on, and with the best success. One would have thought that the moment the foals were let loose in the bush, where they were but seldom seen, that when next brought to the stockyard, which might not be for from fifteen to thirty days, that they would have forgotten their former lesson. Such, however, was not the case, for with *no further second handling* they became daily more familiar, as their instinct strengthened; as *if the results of their schooling had begun only, after it was passed, to dawn upon them.* Except catching them now and then, I took no further trouble with them, than to Rareyfy them, perhaps, three times in their first month, until they became two years off. They were then mouthed, ridden for a fortnight, and turned out. They made no resistance to the breaking whatever, and however fat or flash they might afterwards be, were remarkably quiet to ride, none of them ever trying to get rid of their riders, or showing vice of any description. One of them, which I still have, though he had never been worked, stood the stockwhip without flinching the first time I tried him, though fat and in high spirits at the time, neither did the rifle fired from his back a few days afterwards, cause him the least uneasiness. In fact, I never saw horses more effectually or so easily broken.

If these results, of which I feel no doubt, can be obtained, it appears that a great deal more may be effected by the breeder than has heretofore been found practicable. For if the treatment of which I made use, produced such remarkable results, that which I have just recommended should have a more decided effect. The conclusion to which I have come, and which I wish to lay before my reader, is, that one man would be able to prepare 100 horses for sale in five years, taking charge of, and beginning to Rareyfy them when foals, and carrying out some such plan as I have described. But let us limit him to 70 instead of 100 horses, and put his wages and keep at £70 a year. This should enable the breeder to dispose of his 100 horses, perfectly broken and made, as three, four, and five year olds, the expense on each horse for so breaking him being just £5. Were they truly well-bred, handsome horses, the lowest price at which I can estimate here is £30 a head;

in India, I believe, they would be worth a £100 a head; the lowest of which figures would handsomely remunerate the breeder. The employment of black boys in some districts, would, of course, still further reduce this figure, as they are excellent horsemen, and require no wages.

Besides the facilities for taming and quieting the horse, which result from the application of Rarey's system, another very important advantage may be derived from it. I allude to the habit of lying down. On this point the Arabs are much more particular than we are. "*Faites peu de cas,*" says Daumas, "*d'un cheval qui ne se couche point.*" On a pretty close observation of the habits of the horse when in the East, I came to attribute a great part of his endurance, as well as the extraordinary soundness of his legs and feet, to his great habit of lying down. Spending but a very small portion of his time in eating, he is constantly lying down on the sand where he is picketed. Some few Australian horses which I have had were remarkable in this way, and possessed capabilities for work above others of similar figure, which I could attribute to no other cause. Some horses, too, have a vicious disinclination to lay down. I have known one that used to fall down in his stable whilst asleep, too or three times a week.

Stonehenge, I notice, has a great dislike both to

Rarey and his system. He hardly allows that it was new to the public, grumbles because a racing mare on which he operated never ran well after, and still worse than this, that the Yankee adventurer drew £25,000 from the pockets of the British public. Whether these are sufficient reasons for damning the system, I leave my reader to determine ; but as that writer recommends the practice to cavalry men, I presume his objections are limited to the racer.

It must be acknowledged that Rareyfying has not taken root in Australia, simply because it has been ill applied, and too much was expected from it. With confirmed buck-jumpers it is perfectly useless, as far as a cure is sought; but with any colts its action is, I believe, unfailing and irresistible. Practically, however, the small outlay at which it enables the breeder to oring his horse to market fully broken, is its great advantage to the Australian. It also favours early breaking, without which there can be no perfection. "*Cet exercice*," says Daumas, of early work, "*convient à tous deux ; l'enfant se fait cavalier, le poulain s'habitue à porter un poids qui est en rapport avec sa force, il apprend à marcher à ne s'effrayer de rien, et c'est ainsi, disent les Arabes, que nous parvenons à n'avoir jamais de chevaux retifs.*" Putting striplings on young colts properly prepared for the saddle, teaches the youth to ride, and does away with the evil of having fractious horses.

ON RACING.

"Les chevaux sont des oiseaux qui n'ont pas d'ailes."

Of whatever the breeder of pure saddle-horses may effect by attention to blood, climate, food, selection, and a choice of site for his stud, racing is the only practical competitive test. To come to this conclusion requires but little consideration, and I will not dwell on the point. It is not less a fact, that distinct and decided characteristics, the result of rewarded performances, or prizes for excellence in any way, will soon arise and stamp themselves on the horse subjected to their influence. In breeding horses, as in most other matters, profit is the great motive power. Thoroughbred horses are bred for what? As the actual beau-ideal of the species? As the most useful, or the soundest, or stoutest of horses? As valuable to the community, or advantageous to any but their owners? Not by any means. The gentlemen of the Turf are not suspected of any such views. They breed their horses to race and win. It is natural to suppose they would do so. It would be folly to expect otherwise. The qualities sought by breeders have reference to this point wholly and solely. Describe the races which will be in vogue here for a hundred years, and I will foretel you the sort of horses which will result from them. Saddle-horses, and the institution of races, are vitally connected.

It is not my intention to enter upon a discussion of English racing, a matter already fully treated of by many abler pens than mine. It is, in fact, a subject of which I claim no special knowledge. Had I been professionlly (if I may use the word) addicted to that sport, or even a too ardent amateur of the Turf, I should, no doubt, have become more or less imbued with those one-sided opinions on the subject, which I have noticed in its votaries, and as a *particeps criminis* but an unfair judge of its results. But, whilst I have seen a good deal of Australian racing, as well as something of the principal English and Irish meetings, I have still always remained a mere outsider and a non-betting spectator, just looking on quietly, seeing what was to be seen, hearing what was to be heard, and endeavouring to estimate this horse-trial at what it was worth. Having thus watched the results of racing on a very fair field for observation, and read in the pages of Stonehenge, Nimrod, and many others,—all at heart special pleaders for the Turf,—their deliberate records of its history and results, or, in other words, its failure as a means of producing good saddle-horses, I cannot but acknowledge that I have been gradually losing all interest in this pastime. Indeed, the mere fact of having come to consider our races as of no more real import to saddle-horse breeding than are billiards or bull-baiting, has necessarily deprived it of all serious interest to a non-racing spectator.

I have no intention of re-producing in this chapter the facts connected with this subject, which I have already laid before my reader in treating of the English horse. If I have not there adduced facts and testimonies strong enough to convince him of the errors of our system, or of the unsoundness, want of stoutness, the general frailty, and progressive decadence both of our racers and saddle-horses resulting from it, I am afraid that I must consider his prepossessions as impregnable, and that nothing which can be added will bring conviction to him. And yet on what score the evidence adduced can be rejected, by any one open to conviction, I am at a loss to understand.

On the one hand, we have the testimony of many adequate and credible witnesses, as Abd-el-Kader, Daumas, Nolan, Shakespeare, and Layard, and a host of travellers between whom the idea of collusion is impossible, to prove in the Eastern saddle-horse a degree of stoutness, soundness, abstemiousness, vigour, health and durability, which are quite unequalled in England. Abundant instances of the existence of such qualities are recorded by them, some of which I have laid before my reader. On the other hand, the boldest stopping short at assertion, no instances of perfect or good saddle-horses existing in England, but as a rare exception, are, as far as I know, attempted to be brought forward by any one; but on the contrary, whilst the (in this respect) ill-informed horse-breeder believes that he is producing an animal of unmatched excellence, the really instructed apologist for the English horse is forced reluctantly to admit the inferiority, the daily increasing inferiority, of the saddle-horse of Great Britain.

Whilst ascribing this decadence, as all who admit it do in great part, to the pernicious tendency of our race-course, there are, it must be borne in mind, a large number of individuals who would dread and struggle against any departure from the present system, however injurious it may be to the public. To some, a change would result in pecuniary loss or ruin. Many who might be convinced, would, nevertheless, dislike to acknowledge that their cherished bubble had burst; that they must breed new horses, consign the old exploded racing calendar to the flames, and unlearn at leisure their now discarded science. These persons would be well backed up by many to whom new ideas are a hard crop, and their culture a bitter drudgery.

There may be reasons why, not being particularly interested in this subject I might have kept my convictions to myself; and I should have done so, had I thought any time would be more fit than the present for improvement. For as yet, in a new land, the many tangled interests, which make abuse too often sacred in older communities, have not yet grown up amongst us. We are as yet in the position most favorable to change, when there is a prospect of amelioration. No doubt the day will come, in due time, when Victoria will produce a fair crop of genuine Victorian-bred absurdities on its own account. In the meantime, why forestal the harvest? Why import foreign absurdities at ruinous prices? Why support this monstrous and palpable folly? Let us not deck with our wreaths the columns of a mouldering temple, or help to prop up a tottering inconsistency. Let us cast aside this antique rubbish—look the matter straight in the face, and call common sense to our councils.

Before proposing a new system of racing, with the definite object of producing *a breed of horses pure for the saddle, in the same sense as the racer is pure for the turf,* let me beg of the reader to recollect the full effects of the one which has worked so much mischief. Not only has it rendered our racers, as Stónehenge and others acknowledge, quite unfit to be, as they were intended to be, the sires of our saddle-horses; not only have they themselves become delicate and incapable of work, but they are absolutely yearly becoming less fit even for the sport for which they are in fact designed. To such an extent has speed been cultivated, so entirely and systematically has every other quality been sacrificed to it, that at last the animal, whilst retaining his speed, has become incapable of bearing uninjured the shocks of his own gallop, even when without a rider: the wheels of the machine have become so crazy that its locomotion has become absolutely dangerous to itself. Hence it is not surprising that, to reach the glittering prizes held out to speed, every ingenuity has been set agoing to spare the feeble failing legs, and train the racer with the minimum of work. To assist this purpose, the Turkish bath has been, as we have seen, brought in, to reduce the fat and purge the humours which have heretofore been worked off by sweating gallops, which the racer can now no longer support.

This infirmity of the racing machine, has not, however, escaped the observation of sporting men in England, and the subject has been brought before the House of Commons. As we have seen, both Stonehenge and Nolan have proposed a government interference, and other remedies for the evil, though, in my opinion, but inefficient ones. In fine, a remedy is rarely sought in such matters, until long after an evil has been discovered; and it may be said that, whilst the racer of England is as fast for a short distance as ever he was, he has notoriously deteriorated in stoutness, soundness, and all other useful qualities, and has quite failed as regards the object for which he was originally cultivated.

In proposing a new system of racing to the consideration of my fellow-countrymen in Australia, the object I have in view is the creation of a pure and perfect saddle-horse. If we are ever to do this, we must, as far as we can, so regulate the practices of our Turf *as to make our races a test of such qualities as are indispensable in the sort of animal we require.* There are, it is true, some very essential points in a good hack, which, practically, can hardly be tested on a race-course; but a great point would be gained, if we could institute races which should fully test and render compulsory on the Turf, the principal characteristics of a good hack, which are, soundness, stoutness, endurance, vigorous health, ability to carry weight, and finally, speed *after considerable fatigue has been supported.* Such speed as is witnessed at a race, is, perhaps, never required in a hack: the race which I shall propose, will be seen rather to be a test of *what remains in the horse, after his work*, than mere fleetness of foot. As matters at present stand, if the two horses *A* and *B* start to run a mile and a distance, *A* will win because he is the faster horse of the two, *not absolutely*, but under the purely adventitious conditions of the race. Then *A*, having won his Derby, becomes the horse of his year. To his owner, at least, he brings renown and fortune; to himself may succeed, and often does, a ruined frame and a breakdown. And yet his name is entered in the lists of fame; though ruined for the Turf, he retires with honor to the Stud, there to propagate his decrepitude, and entail his inutility. Whilst his antagonist *B*, damned for losing by a length, perhaps vigorous, sound, unscathed by the ordeal through which he has past, eminent in every *useful* qualification, is heard no more of. If, in a trial of soundness and stoutness, *A* and *B* were found equal, the superior speed for the mile and a distance, would, in itself (and *as a test of remaining power*), be a valuable quality superadded, and deservedly carrying off the victory.

Soundness and stoutness are the first necessities, and then speed: in the union and pre-eminence of these qualities lies the perfection of the saddle-horse. To cultivate such an animal here, the assistance of

Government would be essential. The advisability and propriety of such assistance I will consider presently. In the mean time, these are the measures which I propose for the perfection of the saddle-horse, in connection with the Turf :—

1st. That for fifteen years our Government should assume the control of our races.

2nd. That our races, as at present conducted, be discountenanced as much as possible by our Government.

3rd. That races, as hereinafter described, be instituted by the Government, courses allotted, and fitting prizes provided for excelling in them.

4th. That, with a view to establishing a breed of pure saddle-horses, a Stud-book be opened, under the direction of the Government, in which shall be inscribed the names of all horses and mares pronounced qualified to be entered in such Stud-book, as well as all such mares and horses as shall be bred pure from this source; or, in other words, whose sires and dams shall have been received into the Stud-book.

5th. That for fifteen years the race-courses be open to all stallions and mares.

6th. That for fifteen years any horse or mare that can comply with the necessary conditions may be inscribed in the Stud-book.

7th. That after the lapse of these fifteen years, no horse

or mare be allowed to be entered for any race, unless the name of such horse or mare be found in the Stud-book, except such horse be an imported Arab.

8th. That no stallion, mare, or gelding, shall be competent to be entered for any race until after having gone through the trial which I have denominated WORK, which shall consist in the performance of certain distances for six consecutive days, terminating on the day of the race itself; coupled with other conditions which will be found in the table annexed.

9th. That no horse or mare shall be entered in the Stud-book—excepting such as claim such entry by descent—unless he or she shall have accomplished the requisite trial on the course, at least twice in three years.

10th. That any horse being distanced in any race, or becoming unsound whilst at the Work or in the Race, be thenceforth disqualified from racing, and consequently for the Stud-book.

11th. That the first horse that passes the winning post shall be the winner of the race, only if he be found to be sound after it. That no stakes shall be awarded after a race, until a reasonable number of days shall have elapsed, in order that any unsoundness may be detected. That the stakes shall be won by the first sound horse that passes the winning post.

12th. That horses, mares, and geldings shall race separately, and only with those of the same age and sex.

13th. That the WORK shall be performed in this way, viz. : that an appointed official shall start at the hour determined on, to ride round the course where the work shall be performed, at a pace as nearly as possible such as is determined on for the occasion, and is indicated in the annexed table. That this official shall be followed at a convenient distance, say of 100 or 200 yards, by a second official, and that all horses entered for the race shall perform the distances allotted to them as work between these two officials, under pain of exclusion from the race ensuing.

14th. That whilst doing the said work, five minutes to breathe be allowed at the end of each hour.

15th. That no gelding be admitted to run unless he be descended from a sire and dam already inscribed in the Stud-book.

16th. That the stakes be allotted to be run for in the ratio of forty to stallions, twenty to mares, and ten to geldings.

17th. That any horse or mare winning as a five, six, and seven year old consecutively, be entitled on the last occasion to double stakes, and be incapacitated from again running.

18th. That the weights named in the table below

be those put upon stallions, and a slight reduction be made in this particular in favour of mares.

19th. That all mares be disqualified from running after seven years old.

20th. That between the work and the race, on the day of the race, half an hour be allowed to intervene.

TABLE SHOWING AMOUNT OF WORK PREVIOUS TO RACES; ALSO, LENGTH OF RACES, WEIGHTS TO BE CARRIED THOUGHOUT BOTH WORK AND RACES, AND RATIO OF STAKES TO BE RUN FOR.

AGE OF HORSES.	WEIGHTS TO BE CARRIED.	EACH DAY'S WORK.	RATE OF WORK PER HOUR	LENGTH OF RACES.	RATIO OF STAKES.
YEARS.	STONE.	MILES.	MILES.	MILES.	
4	9	20	6	2	50
5	10.7 lb.	30	7	$2\frac{1}{2}$	300
6	12	40	7	3	700
7	13	50	8	3	1,000
8	13.7 lb.	50	9	$3\frac{1}{2}$	400
9	14	50	9	4	400

From the above, which does not purport to embrace the minutiæ of the subject, the reader will be able to understand the system of races which I advocate, and which, I believe, would cultivate and test most of the great essential qualities which are required in the saddle-horse. In such races, I believe, none but horses fit for the saddle would excel; and that, after some experience on the subject, the weights and distances, if not found the most advisable, might be so re-adjusted as to cultivate and eventually ensure a horse uniting speed, endurance, and soundness, in the most useful and perfect combination. To bring into use a system such as the one which I am advocating, the interference of the Government would, as I have said, be requisite, and with this subject are connected several considerations, of which the most prominent are the following, viz. :—

Are our horses likely to be extensively ameliorated without any interference on the part of Government? Is the improvement of our horses a subject of sufficient importance to warrant the interference of our Government? Is the temporary interference in horse-breeding a matter of such a nature as a Government can judiciously take in hand to a certain extent? To these questions I will reply as briefly as I am able.

1st. *Are our horses likely to be extensively ameliorated without any interference on the part of Government?* In answer to this question let me ask another. Why should we expect amelioration from a system which has been tried extensively in England, here, and elsewhere, and always proved a failure? Or, is there anything to justify us in the expectation that a change to a new and better system will gradually come about unaided by Government? Let us again judge and be guided by the past. In England, a change in the regulations of the Turf, with a view to the improvement of the horse, has often been sought, and always been overruled. Those most immediately and intimately alive to such change being, of course, those who would suffer from its effects; for the seekers of future and public advantages are but few, and weakly moved; expectants of immediate and personal loss, numerous and energetic. We shall never see our racecourse radically improved by our racing men! Interest and prejudice bind them to things as they are. They have invested in that system. Their merchandize is of value, only whilst it lasts. It is their security.

2nd. *Is the improvement of our horses a subject of sufficient importance to warrant the interference of Government?*

Though I ask the question, I can hardly think it requires an answer. Is any wide-spread industry or interest beneath the consideration of a government, when such interference can be made beneficial? Say, looking to the future, that in a hundred years there will be a million of saddle-horses stabled and at work in New Holland—and, that half the number might do the work as well or better, is not the probability of such a result worthy of the consideration of our Government of to-day? Is not an Indian, an American, and a French, aye, and an English market for our stallions worth looking after? Is not a horse power, and what else do we seek in all ages but power? Is not a good article preferable to a bad one? When industry has evidently and permanently gone astray, shall not Government take it by the hand and lead it back to profitable paths? It is one of the very duties of a government—such acts are not forgotten: as many Frenchmen know that Napoleon was, I may say, the creator of beet-root sugar in France, as that he triumphed at Wagram or Eckmuhl. It is as important, and fit a care for a state, that the acreage, capital, and labour devoted to the cultivation of the saddle-horse be advantageously used, as that any other industry equally extensive should be on a proper footing.

3rd. *Is the temporary interference in horse-breeding a matter of such a nature as a Government can judiciously take in hand?*

First of all, it may be remarked, that horse-breeding has at various times been largely interfered with, and with most excellent results, by the governments of England, France, Russia, Germany, and Persia, and in as marked a manner as that which I recommend. The present Emperor of the French, who is understood to be personally thoroughly conversant with the subject, has in France done more for the horse than any man of our times. In several countries, governments have had to choose between importing and breeding their cavalry horses : fully showing that commerce and the spirit of gain do not always contrive to supply a market, or out-weigh prejudices. Further than this, the interference of the Government in England is even now, and long has been, looked forward to for assistance by many supporters of the Turf, and others interested in the horse : hence, the idea of a government interference is no novelty at least. We legislate for the sheep, the duck, the snipe, and is the horse of less consequence ? Who can tell the amount of money that France has spent abroad in the last fifty years on this item alone, which the Government should have enticed and coerced its citizens to admit into their own pockets ? From this very cause Germany has more than once embargoed the export of horses, of which she was being drained to mount French troopers. It is, I believe, not too much to assert, that no European people have heretofore bred even moderately good horses, where that industry has not met with the *surveillance* of their governments. On this subject, says Stonehenge, "I believe that a government inspection of all horses and mares used for breeding purposes would be a great national good, and I look forward to its establishment, at no distant time, as the only probable means of insuring greater soundness in our breeds of horses. I would not have the liberty of the subject interfered with: let every man breed what he likes; but I would not let him foist the produce on the country as sound, when they are almost sure to go amiss as soon as they are worked."

Though Stonehenge thus advocates the interference of Government, he does so in a manner falling short of that which I have proposed. He would modify, renovate, and patch up a bad and moribund system; I would eradicate it without mercy, and substitute one founded on experience and common sense. He would lay inferior horses under a disability; I would, besides this, absolutely and directly encourage the breeder of good horses by handsome rewards for success, and rob the breeders of greyhounds (for they are not worthy to be called horses) both of their *prestige* and profits. I would evoke a new and national spirit on the subject, and direct to useful ends the aims of our horse breeders. I would cede to the Englishman the honor of saying, "I have a horse that with a feather on his back is unequalled for a mile. He can, when he does not break down, do his mile in 2 minutes 20 seconds, or 1 minute 20 seconds (if you like)." I would enable the Australian to reply, "I ride 14 stone, and never require

ON RACING. 283

to do a mile in 2 minutes 20 seconds, still less in 1 minute 20 seconds. My horse can carry me 170 miles in one day, or 100 miles four days consecutively. He lasts 20 years, is never lame, and can live well where yours would die of hunger. When I want a useless toy, I will buy your thoroughbred; when you require a pure saddle-horse, you will know where to seek him."

And here it may be asked ; can any great, general reasons be advanced why saddle-horses should not be bred amongst us in perfection without some authoritive supervision ? My reply is, that such reasons do exist and very strong ones ; first, the thing to be appreciated must be known. Then again, by crossing the pure saddle-horse a great increase in size may be effected without much real damage to quality becoming apparent to the eye of the ordinary purchaser. To a preverted taste, the cross-bred may even be preferred in the matter of beauty. For this increase of height and *quasi* beauty the breeder realizes an advanced price, which quite satisfies him as to the advisability of the cross. Now in truth this increase of size, which is paid for by his buyer, is obtained simply by the addition of elements unfitted and damaging to the animal he purchases. The more his size is increased, the greater the addition of incongruous elements, of elements interfering with, damaging, unsuited to the purpose for which he is supposed to be bred. It is the gold nugget alloyed with baser metal. Like the coffee-measure heaped up with chicory—he is a deception—a hoax—an adulteration—and sufferable only to preverted tastes.

There is one other assistance which the Government might be asked to lend to the undertaking, I mean the purchase of half a dozen *high caste* Arab sires. It is true they would not, theoretically, be indispensable to the undertaking, though they would certainly wonderfully hasten and facilitate it. But the attainment of such horses would be beyond the means of private persons, *for money alone will not purchase first-class Arabs of high caste.* The attainment of one such horse would be an affair of some difficulty, and the purchaser should unite something of the character of an ambassador to that of a merchant. In order to provide prizes for the races, in any of our colonies where such a system as this should be carried out, as well as to defray the cost of the small staff necessary to manage the affairs of the Turf, I think a tax on all horses in the district would be a very proper measure. It would press lightly on the public, and tend highly to their advantage. It would help to put a stop to the excessive production of horses, which, at the price of £2 and £4 per head, now crowd our market, and uselessly consume our pastures; of horses which are useless to the state, and yet stand in the way of the breeder of better animals. Still further, and in addition to the registration proposed by Stonehenge, to stop this evil, I would advocate a yearly tax, say of £50 a head, on all stallions over one year old, excepting the pure saddle-horse sires. The proceeds of such taxes, properly applied, would, I think, soon lead to a marked improvement in the quality of our horses.

In the foregoing proposal to institute races of a more useful tendency than those now in vogue, it must not be supposed that I have attempted to do more than suggest such general measures as appear calculated to ensure the result which I have in view. They will, I dare say, meet with the same encouragement from racing men that Torrens's Act did from the lawyers. If they are honored with any notice, a hundred objections may be raised to the proposed innovation. Some will attack the letter, and avoid a discussion of the principle. Some will condemn without considering my proposal. By some I shall be told that no horses legs can stand the trial to which I would submit them. "Fifty miles a day," I think I hear some one say, "for a week, carrying 13 stone, winding up with a $3\frac{1}{2}$ miles gallop! It would break down any horse." I know as well as I can be told, that this sort of trial will not suit the legs of one English thoroughbred out of five hundred. I know this, as well as I know the folly of cultivating such horses. I also know that there are breeds of horses which can do all I propose with ease, and more, and that if the public encourage such here, we shall gradually succeed in replacing the wretched animal we now have to ride with real horses.

Neither is a long race, and one which tries the endurance of the horse much more severely than the one which I advocate, without example, as the following passage from Youatt, when speaking of the Persian horse, will prove: "My curiosity," says the author from whom he quotes, "was fully on the spur to see the racers, which I could not doubt must have been chosen from the best in the nation, to exhibit the perfection of its breed before the sovereign. The rival horses were divided into three sets, in order to lengthen the amusement. They had been in training several weeks, going over the ground very often in that time; and when I did see them, I found so much pains had been taken to sweat and reduce their weight, that their bones were nearly cutting the skin. The distance marked for the race was a stretch of four-and-twenty miles; and that his majesty might not have to wait when he reached the field, the horses had set forward long before, by three divisions, from the starting point (a short interval of time passing between each set), so that they might begin to come in a few moments after the king had taken his seat. The different divisions arrived in regular order at the goal, but all so fatigued and exhausted, that their former boasted fleetness hardly exceeded a moderate canter when they passed before the royal eyes." These races were evidently not an idle sport merely, but a test of the powers of the horses. The King of Persia *bred all the horses in his kingdom,* and this was the means he took to test the speed, soundness, and endurance of the sires he used.

In concluding this chapter I must beg of my reader to judge it has a whole, and of the principles which it embodies rather than piecemeal: as such only I offer it to his criticism. Let the reader bear in mind, before he quarrels with my weights and distances, that the more exacting we are of perfection, the greater will be our success; and that when a trial of utility, as on our course at present, has been reduced to say six or eight stone for one and a half miles, a great and evident prejudice has to be combated. For myself, I seek for no shadows : I advocate a legislative interference on the subject, and such races as shall be the test of good saddle-horses, and encouragement to the breeders of them : such races as none but a good saddle-horse can shine in : such races as would eventually raise our saddle-horses to the maxim of perfection. I seek for the discouragement and ruin of the present race-course: that costly and useless folly, which has, as a *means of producing first-class saddle-horses,*—regard had to the money lavished upon it,—proved a *complete failure wherever it has been tried,* whether in England or Australia. Let it be admitted, if you please, that, as in England, better horses are thus bred than in France or Germany, or elsewhere, where little Eastern blood has been used, or trouble or expense incurred in the matter: let this be admitted, for it does not debar the conclusion that a smaller outlay, coupled with judicious management, would have effected infinitely more satisfactory results.

ON RIDING LONG DISTANCES.

"As it has fallen to my lot, from long practice, to have become experienced in horsemanship, so do I wish to point out to my younger friends, how I think they can use their horses most properly."—*Xenophon.*

On the subject of how horses should be ridden and treated on long journeys, there exists amongst people a considerable variety of opinion. I have more than once heard this matter discussed by bushmen, townsfolk, and cavalry men.

The two points on which persons differ are, as to the proper divisions of long distances into days' journeys, and secondly, the pace or speed at which it is most advantageous to the horse—or most economical of his strength—that he should be ridden. In neither Youatt nor Stonehenge do I find any notice of either of these points. Captain Nolan in his chapter " The March," has the following observations : " In the campaigns of the last great war in

Europe, it was no uncommon occurrence to see cavalry arrive on the field quite crippled, having lost half their numbers before a shot had been fired, the remaining horses being in such wretched condition as to be totally unfit for active service. . . . The pace" (on the march) "should be a slow trot, about six or seven miles an hour, the men rising in their stirrups, and walking the horses up and down hill. The horses get in earlier to their food, they are groomed and better looked after, and have more time to refresh.

"If you walk all the way, the horses are kept saddled many hours more than is necessary, the men, get tired, sit unsteady in their saddles, and the horses get sore backs.

"The crawling kind of march really fatigues men and horses much more than a march at a smart trot. *Let any man ride a journey of twelve or fifteen miles* at a walk, without ever breaking into a trot or canter, and tell us when he dismounts how he feels. The horse is always distressed by being too long under the saddle, even though he stands stock still all the while."

Now in all this, as far as the horse is concerned, except the last assertion, I quite differ with Captain Nolan, and feel no hesitation in saying that he has had but little practical experience on the subject, or, at all events, has profited very little by such as he may have had. As far as the horseman is concerned, it is, no doubt, less tiresome to canter fifteen miles, than to ride them at a walking pace. At the same time, it is quite common in this country to ride at a walk for twelve hours a day, for ten, and even twenty, consecutive days, without the horseman feeling at all distressed. I have often done so for a month together, and after a few days became quite inured to the pace, and looked for nothing else. But after all, on a long journey, it is the horse that must be chiefly considered; for then the well-being of the quadruped becomes the first interest of his rider. Now Nolan would have his cavalry horses, weighted with twenty stone, *trot to save their strength and condition*, and lest dragoons, who had ridden 15 miles at a walk, should be too much exhausted to clean and feed them! If such is the case, these dragoons must be very washy dogs! How would this system answer with the infantry soldier, with or without his 60 pounds weight of ammunition, accoutrements, &c., about him? Would it suit him to run a few miles, that his march might be sooner over, his load more speedily removed from his back, and his food at once placed before him? Would running economize his strength? Would it not suit him better to rest a while when half way, and so recruit himself a little? Does any living thing hasten to avoid fatigue?

The Arabs are accustomed to long journeys, and

say of the European style of riding them, as we are told by Daumas, "You Christians trot your horses. So do we, now and then, when not at work, to keep our horses in wind. In time of war, we only use the walk. If we are not in a hurry, a walk is fast enough. It is the lasting gallop. If we are in danger, then the gallop saves our heads."

Experience has taught me the truth of our old adage, "It is the pace that kills." The principle is a general one, and the engineer knows that the steamer or locomotive whose pace is to be doubled, must be supplied with a great deal more than a double allowance of fuel. That the wear and tear of the machine will be much more than doubled. Speed is dear, and those who would have it must pay its cost. What you exact of your horse in speed, must be paid in duration and condition. Thus a horse can walk sixty miles, but he cannot trot or canter that distance.

On a long journey, progress quietly; never remain two hours on the saddle without dismounting for a few moments to ease your horse's back and legs: *you will find it, beyond comparison, easier to save condition than to replace it.*

Of all paces, that at which the horse can accomplish the greatest distance, that at which he works with the least cost of toil and condition, is undoubtedly the very ungainly and disagreeable one of the jog-trot, which generally covers five miles within the hour. This is well known in the Sahara. It combines the advantages of being tolerably speedy, soon bringing the work to a conclusion, with a pace, easy to the animal. Whether it would suit horses carrying twenty stone I cannot say.

Two miles at a trot are as exhaustive as three at a walk. Walking out, at the top of the animal's speed, is scarcely less fatiguing than a gentle trot. To man or beast it is more tiring to walk really hard than to jog easily; hence runners on foot adopt the latter pace: hence, the horse pushed at a walk, tries to spare himself by breaking into a jog, and when getting weary, he must either jog or relax his walk; the horse pressed at a trot is difficult to restrain from consulting his ease in a canter. To ride at the top of any pace is uneconomical of strength.

In this matter of pace, it is usually found that those who cannot by any possibility know anything on the subject, are the most opinionated, and provided with the wisest maxims: it is so easy to theorize on such matters, when the ignorance of the theorist is perfect! No one need plume himself on his knowledge on this head, because he can ride a fat horse 300 or 400 miles, at the rate of thirty miles a day; but double the whole distance without increasing the length of the day's journey, and those not accustomed to such things, left to their own guidance, invariably shut up their horses long before the finish. For a short journey you may ride your horse as you please. The proper and the improper system prove themselves on a long journey.

The English, as a people, know little of riding horses long distances. Until they became a great commercial people, their roads were all but impracticable, and the necessity for much change of place trifling and exceptional. Since the great extension of English commerce, the hack has become too slow to keep pace with the requirements of the man of business, and the coach first and the iron-horse since have completely supplanted him: hence their books make no mention of this subject.

We now come to the consideration of the stages into which a long journey can be most properly divided, and I will here call the attention of the reader to one of the anomalies of the horse: it consists in the dissimilarity of his performance and capabilities from those of man and other animals: thus, the horse favourably ridden will, for a day, leave far behind the best pedestrian. An average colonial horse will carry his rider 100 miles in twelve or fourteen hours—a right good man will have enough to do with fifty miles during the same period; and yet, whilst the pedestrian would have little difficulty in doing the half of this distance say twenty-five miles, daily, for many consecutive days, the same horse would certainly last but a very short time at fifty miles a day, or the half of his maximum. This remark applies more to the half-bred saddle-horse than to the Arab, or the hack of the basest description. The latter may work quietly for a few days, but he cannot perform a long day's work—it kills him; the former, the Arab, is more akin to man in the nature of his powers—he can make a long stretch for one day, and work hard every day. This is to be accounted for in two ways: he is pure for the saddle—his formation, instincts, and powers are especially calculated and cultivated for that purpose. Besides this, he has great *resignation*: work neither irritates nor excites him: it agrees with his nature—he is adapted to the saddle. His day's work over, his frugal supper is soon eaten—no long hours of gormandizing detain him from rest;—he is soon extended on the sand, and asleep, whence he rises in the morning fit for fresh toil. The Arabs, Daumas tells us, say, "*Faites peu de cas d'un cheval qui ne se couche point.*" Our horses are much wanting in this respect.

I will now ask my reader how he would apportion his daily distances, if he had to direct the march of a body of cavalry, which had to perform 1,500 miles in 100 days. The task should not be a very trying one, and is yet capable of being managed in this point alone with widely different results. The way in which such things are usually managed is to divide equally the 105 miles, which would be the weekly proportion of the task, between the six first days in the week, and rest on the Sunday; and this constant harassing, for it could not be called work, would leave the majority of the horses quite unfit for active service at its termination: for the saddle-horse disimproves when unaccustomed to his water, forage, and place of rest, often refusing to lie down: for to lie down is his greatest mark of confidence. It would, in fact, be a most unfit way of performing the journey. I should lay the work out something in this way, presuming accommodation, forage, &c., to be equal everywhere:—

DAY'S WORK.	MILES PER DAY.	TOTAL MILES.	DAY'S REST.	MILES.	DAY'S WORK.	DAY'S REST.
4 at	20 =	80	1			
4 ,,	15 =	60	2			
2 ,,	25 =	50	1			
2 ,,	30 =	60	2			
5 ,,	30 =	150	4	520	21	17
4 ,,	30 =	120	7			
21		520	17			
6 ,,	30 =	180	3			
5 ,,	30 =	150	0			
2 ,,	10 =	20	7	350	13	10
13		350	10			

DAY'S WORK.	MILES PER DAY.		TOTAL MILES.	DAY'S REST.
4	at	30 =	120	2
4	,,	30 =	120	3
4	,,	30 =	120	7
12			360	12

MILES.	DAY'S WORK.	DAY'S REST.
360	12	12

4	,,	30 =	120	3
4	,,	30 =	120	2
2	,,	15 =	30	0
10			270	5

270	10	5
1500	56	44

Horses thus travelled would never have been worked more than five consecutive days, on an average much less—would have been gradually accustomed to their work—would have had many rests, and three of a week each. They would have become inured to carrying their riders thirty miles four days consecutively, and should end their march in good condition—fit to *commence* a campaign.

I have concluded, gentle reader, my self-imposed task. Let me finish with the hope that what I have offered to your consideration, as it is certainly the result of conviction, may meet with impartial consideration, awaken some interest on this subject amongst us Australians, and lead to some beneficial results. It is of little consequence that what I have proposed is new if it be reasonable, advantageous, and practicable. What if it be against the practices of England? It accords with usages still more ancient and general. But that aside: old countries are not necessarily dowered with a perfection of knowledge. Many new things, and simple and useful, have come to us from the barbarous and the uncivilized. From young America we lately learned important facts connected with the very animal I now writ of. Rarey, fresh from his studies in the backwoods, electrified the horsey and horse-learned men of Europe with the new practical thing he had to teach. Though an illiterate man and guiltless of Leicestershire and Epsom lore, he had weighed and appreciated the horse for himself. His conclusion was beside old experience and opinion; but he demonstrated that it was correct, and made easy what had seemed impossible. For partial amelioration and progressive bettering of an industry or profession, look to its followers. Legatees of knowledge, they would again bequeath, improved perhaps, but still their own inheritance. Their minds, conservative in this, stop with additions, corrections, and prunings, and run ever in the worn groove. Radical improvement—the new—more often comes from the amateur unprofessional observer, who has little prejudice, knowledge enough, and an unfettered judgment.

Be this as it may, of one thing my reader may be certain, that my views and opinions, correct or otherwise, have been nursed in no holiday-school; neither are they a *réchauffé* of the mouldy dogmas of turf or stable. They are, indeed, the results of much sweat and long journeys, weary miles, painful roads, and worn-out spurs: intimacy with many who have lived in the saddle :—and experience of the horses of many countries.

If it seems that I have spoken too warmly or too confidently on my subject, differing with so many, I can only plead in excuse, together with my little practice with the pen, the conviction that I am right, the wish to persuade, and the hope of being useful.

FINIS.